T0397342

Gas Hydrate in Carbon Capture, Transportation and Storage

Technological, Economic, and Environmental Aspects

Editors

Bhajan Lal

Associate Professor
Chemical Engineering Department
Universiti Teknologi PETRONAS
Seri Iskandar, Perak, Malaysia

Anipeddi Manjusha

Chemical Engineering Department
Universiti Teknologi PETRONAS
Seri Iskandar, Perak, Malaysia

CRC Press is an imprint of the
Taylor & Francis Group, an **informa** business
A SCIENCE PUBLISHERS BOOK

Cover credit: Universiti Teknologi PETRONAS, Malaysia.

First edition published 2025
by CRC Press
2385 NW Executive Center Drive, Suite 320, Boca Raton FL 33431

and by CRC Press
4 Park Square, Milton Park, Abingdon, Oxon, OX14 4RN

CRC Press is an imprint of Taylor & Francis Group, LLC

Library of Congress Cataloging-in-Publication Data (applied for)

ISBN: 978-1-032-69206-7 (hbk)
ISBN: 978-1-032-72640-3 (pbk)
ISBN: 978-1-032-72636-6 (ebk)

DOI: 10.1201/9781032726366

Typeset in Palatino Linotype
by Prime Publishing Services

Preface

Over the past few years, the demand for energy has been increasing for economic progress which is mainly being fulfilled by use of fossil fuels. With carbon dioxide constituting a significant portion of anthropogenic greenhouse gas emissions, urgent measures are needed to address its impact on the environment. Despite challenges in transitioning from fossil fuel-based systems to non-fossil alternatives, CCS emerges as a crucial interim solution for reducing CO_2 emissions while fossil fuels remain a substantial energy source CCS involves capturing CO_2 from various industrial sources and securely storing and transporting it in depleted oil and gas fields, saline aquifers, and deep-sea sediments. Various processes have been proposed and implemented for capture of carbon dioxide. The application of gas hydrate formation phenomenon in capturing, storing of carbon dioxide is one of the recent areas attracting attention. Gas hydrate-based carbon capture, storage and transportation is a green process which requires water as a solvent and CO_2 which is abundant in nature. There is a pressing need to have a clearly and well-written textbook that focuses on gas hydrate-based carbon capture, storage and transportation fundamentals, and other industrial aspects. The book covers all aspects of Gas Hydrate technology for carbon capture, transportation, and storage. It also considers the rules and regulations involved in decarbonization, the health and safety aspect for gas hydrate technology, the techno economic analysis for the technology. The authors firmly believe that good understanding of these fundamentals is necessary to analyze or evaluate the performance for any of the existing and known carbon capture techniques including gas hydrate-based CCS processes. Moreover, understanding the fundamentals would allow for critical evaluation of novel schemes or devising new schemes. This book consists of eleven chapters discussing different aspects gas hydrates-based carbon capture as a potential research area to be explored further.

Chapter 1 presents a brief introduction on definition, structure, formation of gas hydrates as well as issues related to hydrate in oil and gas systems.

Chapter 2 focuses on chemical hydrate promoters that can be used for different applications which include thermodynamic and the kinetic promoters.

Chapter 3 focuses on the transportation of the hydrate formed through the pipelines; the different economic models used for the study.

Chapter 4 summarizes the different reactor systems and operation modes used for hydrate formation and the limitation of the process.

Chapter 5 discusses the techno economical aspect of the hydrate-based carbon capture, the cost analysis, model steps involved, future prospectives.

Chapter 6 discusses the basics of carbon dioxide storage, conventional pathways used for the storage. It also describes the storage of hydrates formed in sediments, porous media.

Chapter 7 present the risk assessment, environmental impact assessment of carbon capture. It also discusses various case studies on carbon capture all over the world, different leakages and consequences associated with them.

Chapter 8 provides an in-depth analysis of the key modelling and simulation methods utilized to predict the rate of hydrate formation to optimize CO_2 capture processes via the formation of CO_2 hydrates. It also covers the underlying principles behind hydrate nucleation and growth and how these phenomena can be modelled and simulated to enhance CO_2 capture processes through the formation of CO_2 hydrates.

Chapter 9 explores the burgeoning field of applying ML and AI to CCS, delving into the myriad ways in which these technologies are revolutionizing the landscape of carbon capture and storage. We will elucidate their application in improving the design and operation of CCS facilities, ensuring secure and leak-free storage, and optimizing the overall CCS lifecycle

Chapter 10 discusses the decarbonization initiatives taken up. It describes the different policies in CCS, climate change agreements, the sustainable and circular economy practices for net zero emissions.

Chapter 11 describes the different sources of information gathered in carbon capture, it also discusses the initiatives of different organizations, agencies and the online sources related to this technology.

It is a great pleasure to thank CRC press for providing this opportunity to share this book with the scientific community. This book could not have been completed without the support and contribution of the authors involved. In the end, we would like to seek your cooperation to share your thoughts and feedback in order to improve our future work.

Bhajan Lal
Seri Iskandar, Malaysia

Contents

Chapter 1

Gas Hydrate in Oil and Gas Systems

Baldeep Singh,[1,*] *Bhajan Lal*[2] and *Anipeddi Manjusha*[2]

1.1 Introduction

Gas hydrates are crystalline structures that resemble ice and are characterized by their non-stoichiometric composition. They are created by the entrapment of gas molecules (referred to as guests) within a lattice of hydrogen-bonded water molecules (known as hosts). These hydrates tend to form under conditions of high pressure and low temperature, where host and guest molecules are held together through van der Waals forces. In a typical gas hydrate structure, approximately 85% of the composition consists of water molecules, which are interconnected through hydrogen bonds to create cages that encapsulate the guest molecules. It is worth noting that the guest molecule can exist in either liquid or gas form, although the majority of instances in practical applications and research involve gaseous guests [1–3]. Methane, ethane, propane, carbon dioxide, and natural gas are a few typical examples of guest molecules. Natural gas hydrates, also known as gas hydrates, are ice-like crystalline formations made of methane molecules trapped within a lattice of water molecules. They typically occur in Arctic permafrost regions and deep-sea sediments, where specific pressure and temperature conditions favour their development. Recently, methane hydrates have drawn a lot of attention since they are regarded to be a potentially large supply of natural gas. Methane gas, which is produced by thermogenic or microbiological processes, combines with water at high pressures and

[1] Petroleum Engineering Department, Rajiv Gandhi Institute of Petroleum Technology, India.

[2] Chemical Engineering Department, Center for Carbon Capture, Storage and Utilization (CCUS), Institute of Sustainable Energy, Universiti Teknologi PETRONAS, Bandar Seri Iskandar, Perak, Malaysia.

[*] Corresponding author: sbaldeep752@gmail.com

low temperatures to form these hydrates. The methane gas is enclosed by a stable lattice made of the water molecules in a cage-like structure [4, 7]. These techniques help identify potential hydrate-rich areas and assess the resource potential. However, the extraction of natural gas from hydrates poses significant technical and economic challenges. Currently, there are several extraction methods under investigation, such as depressurization, thermal stimulation, and chemical methods. These methods aim to release the trapped methane gas while maintaining stability and minimizing environmental impacts.

1.1.1 Types of Hydrates

1.1.1.1 Type I Hydrates

Structure and Composition

The structure of type I hydrates is made up of two types of cages: small cages, which are dodecahedra with 12 pentagonal faces, and big cages, which are tetrakaidecahedra with 12 pentagonal faces and 2 hexagonal faces. Forty six water molecules make up these hydrates. Theoretically, the formula is X-5 H_2O, where X stands for the hydrate former and hosts a guest molecule in each cage.

Common Type I Formers

Type I hydrate formers include methane, ethane, carbon dioxide, and hydrogen sulphide. In CH_4, CO_2, and H_2S hydrates, guest molecules can occupy both small and large cages. In contrast, ethane inhabits only large cages.

1.1.1.2 Type II Hydrates

Structure and Composition

Type II hydrates possess a more complex structure than Type I. They are composed of two types of cages: dodecahedra with 12 pentagonal faces (small cages) and hexakaidecahedra with 12 pentagonal faces and 4 hexagonal faces (large cages). Type II hydrates consist of 136 water molecules. The theoretical composition, if guest molecules occupy all cages, is X-5% H_2O. Alternatively, if guests occupy only large cages, the composition is X-17H_2O.

Common Type II Formers

Common type II formers in natural gas include nitrogen, propane, and isobutane. Nitrogen can occupy both large and small cages, while propane and isobutane exclusively inhabit large cages.

1.1.1.3 Type H Hydrates

Structure and Composition

Type H hydrates are less common and always double hydrates, requiring both a small molecule (e.g., methane) and a type H former. They are constructed from three types of cages: dodecahedra with 12 pentagonal faces (small), irregular dodecahedra with 3 square faces, 6 pentagonal faces, and 3 hexagonal faces (medium), and irregular icosahedra with 12 pentagonal faces and 8 hexagonal faces (large). A unit crystal comprises three small, two medium, and one large cage, totalling 34 water molecules.

Type H Formers

Type H formers are typically larger molecules, such as 2-methylbutane, 2,2-dimethylbutane, and others. Two formers are required to create a type H hydrate, with the small molecule occupying smaller cages and the larger type H former exclusively inhabiting large cages. These hydrate types exhibit varying structures, compositions, and formers, each with distinct characteristics and implications for their occurrence in natural gas systems [8–11].

1.1.2 Enhancing Factors for Hydrate Formation

1.1.2.1 Impact of Turbulence

1. *High Velocity*: Hydrate formation becomes more favourable in areas characterized by high fluid velocity. This propensity is particularly prominent in choke valves, where natural gas experiences a significant temperature drop due to the Joule-Thomson effect. Additionally, the narrowing of the valve leads to increased velocity, further promoting hydrate formation.
2. *Agitation Effect*: The phenomenon of hydrate formation is significantly influenced by mixing within pipelines, process vessels, heat exchangers, and other components. This mixing can occur not only due to dedicated mixers but also due to complex routing in the system, all of which contribute to enhancing hydrate formation [12].

1.1.2.2 The Role of Nucleation Sites

Nucleation Site Significance: Nucleation sites play a crucial role in hydrate formation, serving as points where the transition from a fluid to a solid phase is favoured. Examples of nucleation sites include imperfections in pipelines, welding spots, and various pipeline fittings such as elbows, tees, and valves. Furthermore, substances like corrosion byproducts, silt, scale, dirt, and sand are conducive to hydrate nucleation [13].

1.1.2.3 The Influence of Free-Water Free-Water Impact

While not an absolute requirement for hydrate formation, the presence of free water has a substantial enhancing effect. It ensures an ample water supply, increasing the probability of plug formation. Additionally, the water-gas interface serves as a favourable nucleation site for hydrate formation [14].

1.1.2.4 Accumulation of Hydrates and Mitigation Strategies

Hydrates formed in pipelines do not necessarily conglomerate at their point of origin. Instead, they can be transported along with the fluid phase, especially in the liquid state. It is the accumulation of hydrates that often leads to issues within pipelines, including blockages, plugging, and potential damage to equipment [15, 16].

1.1.2.5 Mitigating Hydrate Accumulation through Pigging

One effective approach for dealing with hydrate-related pipeline challenges is the use of "pigging." This process involves the insertion of a specialized tool, commonly referred to as a "pig," into the pipeline. While modern pigs serve a variety of functions, their primary purpose remains pipeline cleaning. These pigs snugly fit within the pipeline and efficiently scrape its interior, effectively removing hydrates and preventing blockages. This proactive method helps maintain the operational integrity and efficiency of the pipeline system [17].

1.1.3 Revolutionizing the Oil and Gas Industry: Gas Hydrate-Driven Innovations

1.1.3.1 Desalination, Concentration, and Separation of Aqueous Solutions

On our planet, a vast amount of water, around 1.5 billion cubic kilometres exists, but a staggering 97% of it is saline and located in seas and oceans. Freshwater resources, which humans can use, are meagre, accounting for less than 3% of the total, with the majority locked in polar glaciers and mountainous regions. This scant freshwater supply is unevenly distributed globally, primarily concentrated in large rivers and lakes.

Efforts to improve freshwater distribution involve creating artificial reservoirs, developing irrigation systems, tapping underground sources like artesian water, and establishing extensive desalination facilities [18–20]. The desalination process is based on the properties of hydrates, enabling:

1. Formation of hydrate structures from pure water molecules and an agent, leaving impurities behind in the solution.
2. Separate crystallization of salt and gas hydrates in a saturated solution, as these two solid phases have significantly different densities, allowing separation by weight.

The desalination of water via the gas hydrate process is similar to freezing on contact. Deaeration, cooling, and solidification are performed on the original solution. When gas hydrates are in contact with a hydrate-forming material, the temperature at which they form is lower than it would be for pure water because salts are present in the solution. In order to acquire fresh water while removing gas pollutants, hydrate crystals are then isolated from the brine, cleaned, and melted. It is adaptive to diverse desalination circumstances, salt content, energy needs, ambient temperature, and crystallizer temperature thanks to the method's flexibility in selecting a hydrate-forming chemical with a clear location on a P-T diagram. The Hydrate Desalination System (HDS) is appropriate for a variety of desalination regimes, including low, medium, and high brine concentration, high brine concentration with dry salt production, and dry salt production alone.

The environmental benefits include preventing brine pollution and recycling fresh water for production [21–23]. Salts can be produced as mixtures or pure components depending on the purity of the phase, crystallized from the multicomponent brine. Environmentally friendly hydrate formers like propane and ozone-friendly freons are recommended. This gas hydrate technology reduces specific expenses by two to three times compared to multi-stage distillation, and 1.5–2 times compared to membrane methods like electrodialysis and reverse osmosis. It is particularly effective for cleaning waste brines with different pollutant compositions on a large scale, with productivity ranging from 500 to over 100,000 m^3/day. The crystal hydrate method offers several advantages over distillation and membrane-based desalination methods.

1.1.4 Gas Hydrate Applications in Transportation and Storage

A proposed method for transporting natural gas involves converting it into a hydrate state and compressing it into solid cylindrical blocks (nonreturnable containers). These blocks are then cooled, transported through pipelines on a gas cushion under natural gas pressure, and melted at the destination for gas release. To enhance transportation efficiency, the centre of mass of each hydrate block is lowered, and a Z-shaped hermetic cuff is used to reduce gas flow between blocks with varying pressures. Blocks can be solid or hollow to reduce hydraulic losses and accommodate other materials like oil, gas condensate, coal, or ore. The blocks have hardened outer surfaces, typically with an ice or hydrate crust. Pressure drops are created at stations using thermal compression of gas between blocks [24, 25]. At station A, hydrate blocks are formed, shaped, and continuously cooled by hydrate converters connected to the transportation pipeline. Station B increases gas pressure and provides continuous cooling, while station C melts hydrate blocks with water to produce compressed gas, fresh water, and coal.

Storing gas in a hydrate state is advantageous due to the high gas-water molar ratio and the exceptionally high gas density. Methane stored in a hydrate lattice has a higher specific density than in its liquid state. This method is most efficient at relatively low pressures, as it can store significantly more gas in a given volume compared to its free state. The benefits of gas storage in a hydrate state stem from the comparison of gas volumes in free and hydrate states at various pressures [26, 27].

1.1.4.1 Efficient Blowout Control with Hydrate Plugs

Hydrate plugs, known for their adhesive and strength properties, offer an efficient solution to manage uncontrollable gas and liquid flows during well blowouts. This method proves particularly effective in situations where conventional approaches are complex or costly, especially when dealing with wells that have damaged well-top equipment or limited access to the site. The Kurzha well, situated on the Arctic Ocean shore, experienced a catastrophic blowout. Various traditional methods were attempted, including drilling parallel wells, drilling into the emergency well bore, and even a nuclear explosion-induced rock shift. Unfortunately, none of these approaches succeeded in sealing the well.

A significant breakthrough occurred with the introduction of an inclined well, Kumzha No. 9, which was used to create a complete hydrate plug within the Kurzha well bore. This involved connecting the two wells at a depth of approximately 1000 metres and injecting cooled drilling mud or water into the flow, resulting in the formation of a hydrate plug in the 800–400 metre interval. This successfully suppressed the blowout. The process was executed by injecting drilling mud at around 0°C into the escaping gas stream, rapidly cooling it during column cutting [28, 31]. Within a mere 40 minutes, a comprehensive hydrate plug was formed, completely covering the well's cross-section. Subsequent actions included the injection of additional water and cementing the well bore, ultimately achieving a secure well seal. This successful well blowout control operation demonstrated the potential of employing hydrate plugs in challenging well blowout scenarios. It showcased their remarkable effectiveness in mitigating such incidents, even when faced with complex conditions.

1.1.5 Extraction Innovations for Methane Hydrates: A Path to Sustainable Energy

Methane hydrates are an enticing energy source with an enormous potential because they consist primarily of methane molecules. These stable, methane-rich molecules present an enormous hydrocarbon energy source. The difficulty, though, is in creating effective, economical, and secure extraction techniques. Different strategies are used by modern extraction procedures, such as thermal stimulation, depressurization, and inhibitor injection. Thermal stimulation, which is frequently accomplished by infusing heat sources, entails increasing the temperature within the hydrate stability zone to cause dissociation. While the inhibitor injection provides specific compounds to change the hydrate equilibrium condition, depressurization attempts to reduce well pressure to promote hydrate dissociation. Depressurization and thermal stimulation appear to be the most promising of these methods, particularly in areas where free gas is found beneath methane hydrates. Despite the difficulties and unknowns that accompany the extraction of methane hydrates, the potential benefits are substantial [32–34]. Accessing this unconventional energy source can have a positive environmental impact due to reduced carbon emissions, making it a subject of significant interest and ongoing research.

1.1.6 Depressurization Process for Natural Gas Hydrate Production

Commercial production of natural gas hydrates can be achieved through methods that alter the thermodynamic conditions within the hydrate stability zone, causing the hydrates to decompose. One such method is depressurization, which involves lowering the pressure within the hydrate-bearing zone to induce hydrate dissociation.

The objective of depressurization is to reduce the pressure immediately beneath the hydrate stability zone, leading to the decomposition of hydrates and the release of methane gas. Depressurization is typically initiated in the section below the Base of Hydrate Stability (BSR) through pumping, especially within the underlying free gas zone. The process relies on the Earth's interior heat energy to facilitate hydrate dissociation. Horizontal drilling in the underlying free-gas zone is often used, with continuous gas removal sustaining the pressure-induced dissociation of hydrates.

The depressurization method is most suitable for deposits where widespread gas exists beneath the hydrate cap. Successful gas production from hydrates using depressurization has been observed primarily in polar regions beneath the permafrost [35, 36].

As the pressure is reduced during the depressurization process, the gas hydrates become unstable which allow gas to flow into the free-gas zone where it can be retrieved [37, 38]. To improve manufacturing efficiency, more analysis and modelling are needed to pinpoint the precise mechanisms and how they interact.

1.1.6.1 Thermal Stimulation for Natural Gas Hydrate Production

The thermal stimulation process is another approach to extract natural gas from gas hydrates. It involves introducing heat directly into the hydrate stability zone. Heat can be provided through various means, including injecting steam, hot water, or heated liquids, or indirectly through electric or sonic methods. This increase in temperature triggers the decomposition of gas hydrates, releasing methane. The thermal stimulation process is energy-efficient, as the heat required for dissociation is relatively low compared to the energy contained in the liberated gas, typically around 6%. Hot fluids like steam or hot water are injected down a wellbore to dissociate the hydrates, and the resulting methane can be brought to the surface through another wellbore [39, 40]. There are two primary modes for implementing the thermal stimulation method: a frontal sweep, similar to steam floods used in heavy oil production, or pumping hot liquid through a vertical fracture between the injection and production wells. However, this method has several drawbacks, including potential energy losses to non-hydrate-bearing strata (thief zones) and the requirement of good porosity in the producing horizon, typically around 15% or more.

While thermal stimulation shows promise for gas hydrate production, its cost-effectiveness depends on various factors. Laboratory studies have explored the heating process using hot water or chamber heating in hydrate specimens and sediments [41–43]. In Japan, where methane hydrates are of particular interest as a new energy resource, depressurization combined with well-wall heating appears to be an economically effective method for production. Research into dissociation data is ongoing to assess efficiency and gain a better understanding of the decomposition process in thermal stimulation-based gas hydrate production.

1.1.7 Chemical Inhibitor Injection for Natural Gas Hydrate Production

This approach involves injecting a liquid inhibitor chemical into the gas hydrate zone to shift the hydrate equilibrium conditions beyond the thermodynamic stability zone, thus prompting hydrate dissociation. One common inhibitor used is methanol, which displaces methane hydrate formation conditions.

While effective, the chemical inhibitor injection process is considered expensive, albeit less so than thermal stimulation. This method also necessitates good porosity within the gas hydrate-bearing layer. The process involves injecting the chemical inhibitor to destabilize the hydrates, leading to methane release [44, 45].

There are two main approaches within the chemical inhibition method:

1. *Chemical Substitution*: This approach involves substituting methane for carbon dioxide, simultaneously recovering methane while sequestering carbon dioxide. Various methods of introducing carbon dioxide to the methane hydrate have been explored, including in the gas phase, liquid phase, or potentially dissolved in circulating pore water. Several analytical techniques, such as Raman spectroscopy have provided valuable insights into this process.
2. *Chemical Injection*: Chemicals like methanol, ethylene glycol, nitrogen, and salt brines are used as inhibitors. These inhibitors depress equilibrium conditions, leading to hydrate dissolution and methane production. The efficiency of these inhibitors is closely tied to the temperature difference imposed during the process. Nitrogen gas combined with heating has shown particular effectiveness.

Inhibitors like alcohols (methanol) and glycols (ethylene glycol) are commonly used. They function based on the same principle, although salts have been explored but have corrosion and vapourization issues [46, 47]. The choice between methanol and ethylene glycol depends on factors such as chemical structure, physical properties, cost analysis, safety limits, environmental considerations, and dehydration capacities.

Successful use of inhibitors involves determining the composition and the necessary amounts of inhibitors, with the requirement that it remains at or below its water dew point. Continuous injection of inhibitors into the system is assumed, and this method is particularly attractive for offshore gas line applications.

While methanol appears to have a lower unit price, the recovery and recycling costs, along with safety risks, make ethylene glycol a more efficient and safer alternative for hydrate prevention in offshore gas lines [48, 49]. Thus, methanol's lower flash point and higher operating costs are significant considerations in the choice between these inhibitors.

1.1.8 CO_2-CH_4 Exchange Method and CO_2-N_2 Gas Mixtures for Hydrate Recovery

The CO_2-CH_4 exchange method presents a promising approach to recover methane and sequester carbon dioxide in marine hydrate reservoirs. This method involves injecting carbon dioxide (CO_2) into these reservoirs to release trapped methane, offering various advantages and complexities. One of the key advantages is the exothermic heat released during CO_2 hydrate formation. This heat can supply the energy required for the endothermic dissociation of methane hydrates, making the process energetically favourable. Additionally, the presence of hydrate-containing sediments helps maintain the structural integrity of the reservoir, providing stability. CO_2 hydrates are also thermodynamically more stable than CH_4 hydrates under relaxed pressure and temperature conditions [50, 51]. Furthermore, this method allows for the sequestration of industrial CO_2 waste, enhancing process economics and reducing water production. It also provides a potential solution for disposal of industrial CO_2 waste, addressing environmental concerns.

However, implementing the CO_2-CH_4 exchange method comes with several challenges. Controlling temperature and pressure during CO_2 injection is crucial to ensure gas hydrate stability. Managing phase transitions of CO_2 during injection into deep-sea environments can be complex. Ensuring direct contact between CO_2 and methane hydrates is essential for effective methane recovery. This requires large hydrate surface areas and permeable sediments. When dealing with reservoirs containing sII methane hydrates, the exchange kinetics can be problematic, adding practical difficulties to the process. Experiments have shown that achieving effective CO_2-methane hydrate exchange in porous media matrices within reservoirs can be challenging.

In certain hydrate reservoirs, such as those in the Gulf of Mexico, sII hydrates are prevalent due to the mixture of thermogenic and biogenic gases. Laboratory investigations have explored the recovery of hydrocarbons from sII hydrates through CO_2 exchange. These experiments involved establishing an arbitrary methane/ethane ratio to generate sII hydrates and subjecting them to pressurized pure CO_2. Remarkably, 99% of methane and ethane were recovered within just 20 minutes, indicating efficient carbon dioxide exchange for both gases [52–54]. The final hydrate composition suggested that a transition from sII to another hydrate structure occurred during the process, facilitating the accommodation of CO_2 into the smaller cavities of sII structure, displacing methane, and ethane effectively.

Injecting CO_2 into methane hydrate reservoirs during the exchange process can lead to CO_2 hydrate formation near injection ports, potentially blocking gas flow. However, injecting CO_2-nitrogen (N_2) gas mixtures can mitigate this issue. These mixtures require more stringent pressure-temperature conditions to form CO_2/N_2 hydrates compared to CO_2 hydrates. Simulations indicate that injecting nitrogen along with CO_2 can help maintain reservoir pressure, reduce methane partial pressure, and preserve reservoir integrity while preventing blockage at injection ports. In laboratory experiments, CO_2-N_2 gas mixtures were used to extract methane from s II hydrates in sand cores. Achieving approximately 26.2% methane replacement with CO_2 was possible in the lower saturation case, while the higher saturation case achieved only 6.5% methane replacement.

These results were achieved after injecting about five pore volumes of gas. However, a thin N-CO_2 hydrate film around the methane hydrates eventually hindered mass transfer rates, suggesting the need for further research on addressing such limitations [53, 54].

1.1.9 *Unveiling the Earth's Methane Hydrate Wealth: Assessment Challenges*

Commercially exploiting methane hydrates presents several challenges, including uncertainty about their abundance, extraction methods, and economic viability. Estimates of global methane hydrate resources vary widely, ranging from at least 100,000 Trillion Cubic Feet (TCF) to possibly much more. These estimates need qualification using terms such as "in-place resources," "technically recoverable resources," and "proved reserves" [55].

1. *In-Place Resources*: These represent estimates of methane hydrate resources without considering technical or economic recoverability. These estimates tend to be the largest.
2. *Technically Recoverable Resources*: This category comprises resources that can be extracted using existing technology but does not factor in economic viability.
3. *Proved Reserves*: Proved reserves are quantities estimated to be recoverable under current economic and operating conditions.

The estimates for methane hydrate resources vary significantly, and caution is necessary when interpreting them. The US Geological Survey has made in-place estimates for US gas hydrate resources but has not established proved reserves yet [56, 57]. Despite the lack of commercial production history, there is growing evidence that some gas hydrate resources could be exploited using existing technology.

The exploration of methane hydrate occurrences relies on seismic exploration to locate anomalous reflectors called "Bottom-Simulating Reflectors" (BSRs). While BSRs are indicative of gas hydrates, they may not be present in all cases, and other indicators like micro faunal groups are considered. Methane hydrates are found in marine sediments hundreds of metres below the seafloor along continental margins and in Arctic permafrost. The deposits can be several hundred metres thick and cover large areas. For example, there is a deposit nearly 500 km long along the Blake Ridge off North Carolina's coast.

Methane hydrate resources are distributed across several regions globally, each with its unique potential and challenges for exploration and future energy extraction. These include:

Gulf of Mexico: The Minerals Management Service conducted an assessment of gas hydrate resources in the Gulf of Mexico. Their report provides a statistical probability of in-place gas hydrate resources, with a mean estimate exceeding 21,000 TCF. Within this estimate, porous and permeable sandstone reservoirs are considered the most promising for future production, accounting for around 6,700 TCF. However, it is essential to recognize that the commercial recoverability of these resources may be limited by current technology [61, 62].

Asian Exploration: Several Asian countries, including Japan, India, China, Malaysia, and Korea, have embarked on extensive exploration projects offshore to assess methane hydrate deposits. For example, Japan's MH21 Research Consortium has conducted significant research and development efforts in this field. China has also initiated a substantial US$100 million program focusing on hydrate exploration. These endeavours demonstrate a growing interest in harnessing methane hydrate resources for energy needs in the Asian region [67, 68].

These regions collectively hold immense potential in terms of methane hydrate resources, although challenges related to commercial extraction, resource validation, and technological advancements must be addressed to effectively utilize these resources for energy production.

1.1.10 *Economics of Gas Hydrates: Balancing Safety, Prevention, and Remediation*

The economics of gas hydrates are multifaceted and have a profound impact on decision-making in the energy industry. While safety, environmental concerns, and regulatory compliance play vital roles in shaping these economics, the overarching motivation for addressing gas hydrates is the financial aspect. This chapter delves into the economics of gas hydrate safety, prevention, and remediation, building upon the following chapters [69–71].

Safety considerations are at the forefront of gas hydrate economics. Although it is possible to quantify the financial cost of accidents through insurance, the ethical responsibility to ensure worker well-being often takes precedence. Oil and gas companies are committed to maintaining accident-free operations, even if it incurs higher costs. The correlation between safety and costs is evident, as accidents often result in significant financial setbacks. Case studies underscore the direct link between safety measures and cost savings, emphasizing the value of stringent safety protocols.

1.1.11 Conclusion

Gas hydrate prevention methods, another economic dimension, encompass considerations of individual cases. The cost of prevention can vary greatly, from installing heating systems around instrument gas-control valves to optimizing process designs. Two primary prevention approaches are chemical injection and heat management. Notably, the widespread use of methanol (MeOH) as a hydrate inhibitor underscores its economic significance, with its consumption projected to rise significantly in deepwater operations [72–75]. Economic studies have thoroughly examined the costs of MeOH injection, considering factors like operating costs, MeOH losses to different phases, and seasonal price fluctuations. Environmental concerns and regulations also add complexity to this economic dimension, potentially rendering MeOH less viable despite its cost-effectiveness.

Furthermore, the economics of gas hydrate remediation come into play when production is halted due to hydrate blockages. Lost production poses a considerable financial challenge, and it is essential to classify it as lost or deferred revenue. Typically, deferred production is calculated at the end of the reservoir's life to account for the time value of money, recognizing that a dollar today is worth more than a dollar in the future. Contracts often stipulate delivery and penalties for non-delivery of hydrocarbons, further impacting the economic aspects of hydrate remediation. Remediation methods can be costly, and time constraints often play a pivotal role in decision-making [76, 77]. Substitution is a common strategy to meet gas supply demands during remediation, but it is limited by the borrowing capacity and may require purchasing gas from the spot market, incurring additional costs.

References

[1] Ruppel, C. D. 2018. Gas hydrate in nature.
[2] Abdulrab Abdulwahab Almashwali, Cornelius B. Bavoh, Bhajan Lal, Siak Foo Khor, Quah Chong Jin and Dzulkarnain Zaini. 2022. Gas hydrate in oil-dominant systems: a review. ACS Omega 7: 27021–27037.
[3] Of Yangtze Oil, A. G. A.-O. J., Gas and undefined. 2017. Gas hydrate—properties, formation and benefits. scirp.org.
[4] Sinquin, A., gas science, T. P.-O. and undefined. 2004. Rheological and flow properties of gas hydrate suspensions. ogst.ifpenergiesnouvelles.fr 59: 41–57.
[5] Nature, E. D. S. J. and undefined. 2003. Fundamental principles and applications of natural gas hydrates. nature.com.
[6] Sloan, E. D. 2010. Natural gas hydrates in flow assurance.
[7] Xiong-Qi Pang, Zhuo-Heng Chen, Cheng-Zao Jia, En-Ze Wang, He-Sheng Shi, Zhuo-Ya Wu et al. 2021. Evaluation and re-understanding of the global natural gas hydrate resources. Pet Sci. 18: 323–338.

[8] Reviews, C. A. K.-C. S. and undefined. 2002. Towards a fundamental understanding of natural gas hydrates. pubs.rsc.org. doi:10.1039/b008672j.

[9] Jinxing, D. A. I., Yunyan, N. I., Huang, S., P. W.-P. E. and undefined. 2017. Genetic Types of Gas Hydrates in China. Elsevier.

[10] Koh, C. A., Sum, A. K., of Natural Gas Science, E. D. S.-J. and undefined. 2012. State of The Art: Natural Gas Hydrates as a Natural Resource. Elsevier.

[11] Byk, S. S., Reviews, V. I. F.-R. C. and undefined. 1968. Gas hydrates. iopscience.iop.org.

[12] Kumar, A., Bhattacharjee, G., Kulkarni, B. D. and Kumar, R. 2015. Role of surfactants in promoting gas hydrate formation. Ind. Eng. Chem. Res. 54: 12217–12232.

[13] Dongliang, L., Hao, P., of Heat, L. D.-I. J., Mass and undefined. 2017. Thermal Conductivity Enhancement of Clathrate Hydrate with Nanoparticles. Elsevier.

[14] Lingjie Sun, Huilian Sun, Chengyang Yuan, Lunxiang Zhang, Lei Yang, Zheng Ling, Jiafei Zhao and Yongchen Song. November 2022. Enhanced Clathrate Hydrate Formation at Ambient Temperatures (287.2 K) and Near Atmospheric Pressure (0.1 MPa): Application to Solidified Natural Gas. Elsevier.

[15] Koh, C. A., Westacott, R. E., Zhang, W., … K. H.-F. P. and undefined. 2002. Mechanisms of Gas Hydrate Formation and Inhibition. Elsevier.

[16] Vysniauskas, A., Science, P. R. B.-C. E. and undefined. 1983. A Kinetic Study of Methane Hydrate Formation. Elsevier.

[17] You, K., Flemings, P. B., Malinverno, A., Collett, T. S. and Darnell, K. 2019. Mechanisms of methane hydrate formation in geological systems. Wiley Online Library 57: 1146–1196.

[18] Yue Qin, Liyan Shang, Zhenbo Lv, Jianyu He, Xu Yang and Zhien Zhang. November 2022. Methane Hydrate Formation in Porous Media: Overview and Perspectives. Elsevier.

[19] Eslamimanesh, A., Mohammadi, A. H., of Chemical …, D. R.-T. J. and undefined. 2012. Application of Gas Hydrate Formation in Separation Processes: A Review of Experimental Studies. Elsevier.

[20] Koh, C. A., Sloan, E. D., review of chemical, A. K. S.-A. and undefined. 2011. Fundamentals and applications of gas hydrates. annualreviews.org 2: 237–257.

[21] Ponnivalavan Babu, Abhishek Nambiar, Tianbiao He, Iftekhar A. Karimi, Ju Dong Lee, Peter Englezos et al. 2018. A review of clathrate hydrate based desalination to strengthen energy–water nexus. ACS Publications 18: 2023.

[22] Xu, H., Khan, M. N., Peters, C. J., Sloan, E. D. and Koh, C. A. 2018. Hydrate-based desalination using cyclopentane hydrates at atmospheric pressure. J. Chem. Eng. Data 63: 1081–1087.

[23] Zhang, J., Chen, S., Mao, N., in Energy, T. H.-F. and undefined. 2022. Progress and Prospect of Hydrate-Based Desalination Technology. Springer 16: 445–459.

[24] Semenov, M. E., Pavelyev, R. S., Stoporev, A. S., Zamriy, A. V., Chernykh, S. P., Viktorova, N. V. et al. 2022. State of the art and prospects for the development of the hydrate-based technology for natural gas storage and transportation (A Review). Petroleum Chemistry 62: 127–140.

[25] Lanlan Jiang, Zucheng Cheng, Shaohua Li, Nan Xu, Huazheng Xu, Jiafei Zhao et al. March 2022. High-Efficiency Gas Storage via Methane-Tetrahydrofuran Hydrate Formation: Insights from Hydrate Structure and Morphological Analyses. Elsevier.

[26] Rehman, A. N., Bavoh, C. B., … R. P.-… S. C. and undefined. 2021. Research Advances, Maturation, and Challenges of Hydrate-Based CO_2 Sequestration in Porous Media. ACS Publications 9: 15075–15108.

[27] 化工学报 H. M. and undefined. 2003. Recent advances in hydrate-based technologies for natural gas storage—a review. cqvip.com.

[28] Khodafarin, R., Osouli, A., Vahedi, M. and Nazari, K. March 2022. Importance of hydrate management: deepwater challenges and looking to the past. academia.edu.

[29] Yousif, M. H., Dunayevsky, V. A., Conference, A. H. H.-S. D. and undefined. 1997. Hydrate plug remediation: options and applications for deep water drilling operations. onepetro.org.
[30] Aman, Z. M., Zerpa, L. E., Koh, C. A., Journal, A. K. S.-S. P. E. and undefined. 2015. Development of a tool to assess hydrate-plug-formation risk in oil-dominant pipelines. onepetro.org.
[31] de Assis, J. V., Mohallem, R., ... S. T.-S. W. and undefined. 2013. Hydrate Remediation During Well Testing Operations in the Deepwater Campos Basin, Brazil. onepetro.org.
[32] Dawe, R. A. and Thomas, S. 2007. A large potential methane source - Natural gas hydrates. Energy Sources, Part A: Recovery, Utilization and Environmental Effects 29: 217–229.
[33] Lin Chen, Hirotoshi Sasaki Tsutomu Watanabe, Junnosuke Okajima, Atsuki Komiya and Shigenao Maruyama. 2017. Production Strategy for Oceanic Methane Hydrate Extraction and Power Generation with Carbon Capture and Storage (CCS). Elsevier.
[34] Chen, L., Feng, Y., Okajima, J., of Natural Gas ..., A. K.-J. and undefined. 2018. Production Behavior and Numerical Analysis for 2017 Methane Hydrate Extraction Test of Shenhu, South China Sea. Elsevier.
[35] Jiafei Zhao, Zihao Zhu, Yongchen Song, Weiguo Liu, Yi Zhang and Dayong Wan. 2015. Analyzing the Process of Gas Production for Natural Gas Hydrate Using Depressurization. Elsevier.
[36] Liang-Guang Tang, Xiao-Sen Li, Zi-Ping Feng, Gang li and Shuanshi Fan. 2007. Control Mechanisms for Gas Hydrate Production by Depressurization in Different Scale Hydrate Reservoirs. ACS Publications 21: 227–233.
[37] Shuang Dong, Mingjun Yang, Mingkun Chen, Jia-nan Zheng and Yongchen Song. Feb 2022. Thermodynamics Analysis and Temperature Response Mechanism During Methane Hydrate Production by Depressurization. Elsevier.
[38] Yizhao Wan, Nengyou Wu, Gaowei Hu, Xin Xin, Guangrong Jin, Changling Liu et al. December 2018. Reservoir Stability in the Process of Natural Gas Hydrate Production by Depressurization in the Shenhu Area of the South China Sea. Elsevier.
[39] Tang, L. G., Xiao, R., Huang, C., Feng, Z. P. and Fan, S. S. 2005. Experimental investigation of production behavior of gas hydrate under thermal stimulation in unconsolidated sediment. Energy and Fuels 19: 2402–2407.
[40] Dallimore, S. R., Satoh, T., Inueh, T., Weatherill, B., Collet, T. S. et al. 2005. Overview of thermal-stimulation production-test results for the JAPEX/JNOC/GSC et al. Mallik 5L-38 gas hydrate production research well. gfzpublic.gfz-potsdam.de.
[41] Wang Hongsheng Dong, Liu Yanzhen, Xin Lv, Yu Liu, Jiafei Zhao et al. August 2017. Evaluation of Thermal Stimulation on Gas Production from Depressurized Methane Hydrate Deposits. Elsevier.
[42] Wang, Yi, Li, Xiao-Sen, Li, Gan, Zhang, Yu, Li, Bo, Chen and Zhao-Yang. 2013. Experimental Investigation into Methane Hydrate Production During Three-Dimensional Thermal Stimulation with Five-Spot Well System. Elsevier.
[43] Jiaqi Wang, Fengxu Han, Siguang Li, Kun Ge, Zhiwen Zheng. November 2020. Investigation of Gas Hydrate Production with Salinity via Depressurization and Thermal Stimulation Methods. Elsevier.
[44] Li, G., Li, X. S., Tang, L. G., fuels, Y. Z.-E. and undefined. 2007. Experimental Investigation of Production Behavior of Methane Hydrate Under Ethylene Glycol Injection in Unconsolidated Sediment. ACS Publications 21: 3388–3393.
[45] Li, G. Danmei Wu, Xiaosen Li, Yu Zhang, Qiunan Lv and Yi Wang. 2017. Experimental investigation into the production behavior of methane hydrate under methanol injection in quartz sand. Energy and Fuels 31: 5411–5418.

[46] Li, G., Danmei Wu, Xiaosen Li, Yu Zhang, Qiunan Lv and Yi Wang. 2017. Experimental Investigation into the Production Behavior of Methane Hydrate Under Methanol Injection in Quartz Sand. ACS Publications.
[47] Liang, Y. Yunpei Liang, Youting Tan, Yongjiang Luo, Yangyang Zhang and Bo Li. March 2020. Progress and Challenges on Gas Production from Natural Gas Hydrate-Bearing Sediment. Elsevier.
[48] Na, S., An Lei, Deng Hui, Sun Jian and Guang Xinjun. 2016. Discussion on natural gas hydrate production technologies. cped.cn.
[49] Kelland, M. A., Svartaas, T. M., Conference, L. D.-O. E., ... and undefined. 1995. Studies on new gas hydrate inhibitors. onepetro.org.
[50] Ray Boswell, David Schoderbek, Timothy S. Collett, Satoshi Ohtsuki, Mark White and Brian J. Anderson. 2016. The Iġnik Sikumi Field Experiment, Alaska North Slope: Design, Operations, and Implications for CO_2–CH_4 Exchange in Gas Hydrate Reservoirs. ACS Publications 31: 140–153.
[51] Tupsakhare, S. S., Energy, M. J. C.-A. & undefined 2019. Efficiency Enhancements in Methane Recovery From Natural Gas Hydrates Using Injection of CO_2/N_2 Gas Mixture Simulating *In-Situ* Combustion. Elsevier.
[52] Nicodemus Abelly, Michael Melckzedeck, Mbega Ramadhani Mgimba, Edwin E. Ngata and Long Nyakilla. 2023. Critical review on carbon dioxide sequestration potentiality in methane hydrate reservoirs via CO_2-CH_4 exchange: experiments, simulations, and pilot test applications. Energy and Fuels. doi:10.1021/ACS.ENERGYFUELS.3C01510.
[53] Dongbin Pan, Xiuping Zhong, Ying Zhu, Lianghao Zhai, Han Zhang, Xitong Li et al. November 2020. CH_4 Recovery and CO_2 Sequestration from Hydrate-Bearing Clayey Sediments via CO_2/N_2 Injection. Elsevier.
[54] Schoderbek, D., Martin, K. L., ... J. H.-O. T. C. A. T. and undefined. 2012. North slope hydrate fieldtrial: CO_2/CH_4 exchange. onepetro.org.
[55] Macdonald, G. J. 1990. The future of methane as an energy resource. Annual Review of Energy 15: 53–83. doi:10.1146/ANNUREV.EG.15.110190.000413.
[56] Kvenvolden, K. A. 1998. A primer on the geological occurrence of gas hydrate. Geol Soc. Spec. Publ. 137: 9–30.
[57] Yoshihiro Konno, Tetsuya Fujii, Akihiko Sato, Koya Akamine, Motoyoshi Naiki, Yoshihiro Masuda et al. 2017. Key findings of the world's first offshore methane hydrate production test off the Coast of Japan: Toward future commercial production. Energy and Fuels 31: 2607–2616.
[58] Kerr, Richard A. 2004. Gas hydrate resource: smaller but sooner. Science; Washington 303(5660): 946–947.
[59] Collett, T. S. 2004. Gas hydrate resources of the United States. researchgate.net.
[60] Ray Boswell, Brian J. Anderson, Scott A. Digert, Gordon Pospisil, Richard Baker and Micaela Weeks. 2011, February. Mount Elbert Gas Hydrate Stratigraphic Test Well, Alaska North Slope: Overview of Scientific and Technical Program. Elsevier.
[61] Milkov, A. V., Geology, R. S.-M. and undefined. 2001. Estimate of Gas Hydrate Resource, Northwestern Gulf of Mexico Continental Slope. Elsevier.
[62] Majumdar, U., Geochemistry, A. E. C., undefined Geophysics and undefined. 2018. The Volume of Gas Hydrate-Bound Gas in the Northern Gulf of Mexico. Wiley Online Library 19: 4313–4328.
[63] Collett, T. S., of Offshore, G. D. G.-I. J., Polar and undefined. 1998. Gas hydrates in the Messoyakha gas field of the West Siberian Basin-a re-examination of the geologic evidence. onepetro.org.
[64] Safronov, A. F., Shits, E. Y., Geology, M. N. G.-R. and undefined. 2010. Formation of Gas Hydrate Deposits in the Siberian Arctic Shelf. Elsevier.

[65] Zheng Rong Chong, She Hern Bryan Yang, Ponnivalavan Babu, Praveen Linga and Xiao-Sen Li. Jan 2016. Review of Natural Gas Hydrates as an Energy Resource: Prospects and Challenges. Elsevier.
[66] Majorowicz, J. A., bulletin, K. G. O.-A. and undefined. 2001. Gas hydrate distribution and volume in Canada. pubs.geoscienceworld.org. doi:10.1306/8626CA9B-173B-11D7-8645000102C1865D.
[67] Wang, J., Wu, S., of Asian earth sciences, Y. Y.-J. and undefined. 2018. Quantifying Gas Hydrate from Microbial Methane in the South China Sea. Elsevier.
[68] Ryo Matsumoto, Byong-Jae Ryu, Sung-Rock Lee, Saulwood Lin Shiguo Wu, Kalachand Sain, Ingo Pecher and Michael Riedel. Nov 2011. Occurrence and Exploration of Gas Hydrate in the Marginal Seas and Continental Margin of the Asia and Oceania Region. Elsevier.
[69] Milkov, A. V., Marine, R. S., geology, P. and undefined. 2002. Economic Geology of Offshore Gas Hydrate Accumulations and Provinces. Elsevier.
[70] Javanmardi, J., Nasrifar, K., … S. H. N.-A. T. and undefined. 2005. Economic Evaluation of Natural Gas Hydrate as an Alternative for Natural Gas Transportation. Elsevier.
[71] Collins, M., Louden, K. E., Milliman, J. D., Posamentier, H. W. and Watts, A. 2005. Economic geology of natural gas hydrate.
[72] Collins, M., Louden, K. E., Milliman, J. D., Posamentier, H. W. and Watts, A. 2006. Energy resource potential of natural gas hydrates. pubs.geoscienceworld.org.
[73] Holder, G. D., Kamath, V. A., of Energy, S. P. G.-A. R. and undefined. 1984. The potential of natural gas hydrates as an energy resource. annualreviews.org 9: 427–472.
[74] World gas conference, H. K.-23rd, undefined Amsterdam and undefined. 2006. Economic study on natural gas transportation with natural gas hydrate (NGH) pellets. members.igu.org.
[75] Moridis, G. J. J., Collett, T. S. S., reservoir evaluation …, M. P.-D. and undefined. 2011. Challenges, uncertainties, and issues facing gas production from gas-hydrate deposits. onepetro.org.
[76] Kumar, A., Veluswamy, H. P., Status, P. J., … F. and undefined. 2022. Gas Hydrates in Man-Made Environments: Applications, Economics, Challenges and Future Directions. Springer 173–192. doi:10.1007/978-981-16-4505-1_9.
[77] Moridis, G. J., Collett, T. S., Ray Boswell, Stephen Hancock, Rutqvist, J. Carlos Santamarina et al. 2013. Gas hydrates as a potential energy source: State of knowledge and challenges. Advanced Biofuels and Bioproducts 9781461433484, 977 1033.

CHAPTER 2
Chemical in Carbon Capture and Storage

*Anipeddi Manjusha** and *Bhajan Lal*

2.1 Introduction

Carbon dioxide capture is a critical strategy in the battle against climate change, entailing the retrieval and confinement of carbon dioxide emissions from diverse sources prior to their release into the atmosphere. This method is designed to counteract the exacerbation of the greenhouse effect and global warming caused by excess CO_2 [1, 21]. Various approaches are deployed for carbon dioxide capture, encompassing post-combustion capture from industrial flue gases, pre-combustion capture from processes involving fossil fuels, and direct air capture from the ambient atmosphere.

The captured CO_2 can be securely stored using two principal techniques: geological storage, which involves injecting CO_2 into subterranean rock formations, and mineralization, wherein CO_2 engages in chemical reactions with minerals to yield stable compounds. These practices play an instrumental role in curbing CO_2 emissions and fulfilling climate objectives. Furthermore, the concept of Carbon Capture and Utilization (CCU) entails transforming the captured CO_2 into valuable commodities such as fuels, chemicals, or construction materials. As we grapple with the pressing need to address climate change, carbon dioxide capture emerges as a pivotal instrument within the global endeavour to foster sustainable and low-carbon energy solutions. The advancement of hydrate-based CO_2 capture technology has shown great promise as a viable and innovative alternative to conventional techniques such as adsorption, absorption, membrane separation, and

Chemical Engineering Department, Center for Carbon Capture, Storage and Utilization (CCUS), Institute of Sustainable Energy, Universiti Teknologi PETRONAS, Bandar Seri Iskandar, Perak, Malaysia.
* Corresponding author: anipeddi_22009697@utp.edu.my

cryogenic distillation. As a relatively new field, researchers are actively conducting extensive studies to better understand the thermodynamic and kinetic aspects of hydrate formation, as well as investigating the effects of various inhibitor and promoter media on the process. CO_2 gas hydrates possess remarkable thermodynamic stability, ensuring that once formed, they can persist for extended periods as a secure and reliable medium for CO_2 storage. The crystalline structure of gas hydrates effectively traps CO_2 molecules, immobilizing them and preventing their release into the atmosphere. This confinement minimizes the risk of CO_2 leakage, a common concern with conventional storage methods. The slow rate of CO_2 release from gas hydrates ensures long-term storage effectiveness, significantly reducing the possibility of unexpected CO_2 leaks and ensuring that the sequestered CO_2 remains safely contained within the hydrate structure for prolonged periods. Currently, direct injection of CO_2 into the ocean is not under active exploration, primarily due to concerns about potential negative impacts on the marine ecosystem and marine life. Field investigations in this area are limited as well. Gas hydrates share several physical characteristics with ice, but they possess greater toughness and a thermal conductivity comparable to liquid water. Compared to traditional CO_2 sequestration methods, CO_2 gas hydrates offer a significantly larger storage capacity. These hydrate-based systems show tremendous potential for applications and could play a significant role in mitigating greenhouse gas emissions through carbon dioxide capture, separation, and export. CO_2 exhibits the lowest hydrate-forming pressure when compared to other components present in flue gas. To separate CO_2 from the other gases, an effective method involves initially creating a solid hydrate phase with a high CO_2 content. After isolating this hydrate phase from the gaseous phase, the subsequent dissociation of the hydrates allows for the recovery of CO_2, which is significantly more concentrated than the original feed. Research indicates that the CO_2 concentration in the hydrate phase is at least four times higher than that in the gas phase. CO_2 exhibits a preference for occupying the cavities at lower pressures compared to gases like N_2 and CH_4.

Despite the demanding thermodynamic and kinetic conditions required to induce hydrate formation, the method continues to attract significant research interest globally. Over the years, the primary focus of research in this area has been on enhancing the hydrate formation process, increasing the rate of formation and gas storage capacity, reducing phase equilibrium conditions for hydrate formation, and minimizing the induction time. Researchers worldwide are dedicated to refining and optimizing the hydrate formation method to unlock its full potential for various practical applications.

2.2 Thermodynamic Promoters

Thermodynamic promoters improve the thermodynamic circumstances that support the production and stability of gas hydrates. They function by raising the pressure necessary for hydrate formation while simultaneously reducing the temperature. Thermodynamic promoters can change these circumstances to change the equilibrium phase behaviour, which makes it simpler for gas particles to get trapped inside the water lattice. THPs find their primary use in either reducing the pressure at which hydrate formation takes place or increasing the temperature required for hydrate formation. Additionally, substances such as propane (C_3H_8), neohexene, acetone, and methylcyclohexane have been investigated. Nevertheless, the broad practical application of these compounds faces challenges due to their high volatility.

Tetrahydrofuran (THF): Tetrahydrofuran (THF) is a frequently used thermodynamic hydrate promoter (THP) that can form its own hydrate at ambient pressure, under conditions around 277.5 K and ambient pressure. The THF hydrate takes the form of sII hydrate with a molar composition of 1 THF molecule to 17 water molecules, filling the large 51200 cages. In the presence of (THF), the hydrate formation behaviour changes for all gas components in flue gas. Although CO_2 typically forms SI hydrates when combined with water, the presence of THF triggers the creation of sII hydrates. In these sII hydrates, CO_2 competes for both the small cages, where it competes with N_2 or H_2, and the large cages, where it competes with THF [71, 72, 19, 81]. When THF molecules are present within the spacious $5^{12}6^4$ cavities of sII hydrates, it diminishes the accessibility of these cavities for CO_2 and other gases. As a result, elevated THF concentrations can cause a reduction in both CO_2 recovery and separation efficiency. Studies indicate that an ideal THF concentration of one mole percent (1 mol%) proves to be effective in separating CO_2 from CO_2/N_2 and CO_2/H_2 systems. [26, 38] examined the hydrate formation process for a CO_2/H_2 mixture system in the presence of THF. They discovered an optimal concentration of 1 mol% THF that led to a significantly improved hydrate formation rate. Below this optimum concentration, the formation rate increased with higher THF concentrations.

Tetra-n-butyl ammonium bromide (TBAB): Another extensively researched semiclathrate promoter is tetrabutylammonium bromide (TBAB), which has shown significant promise for various applications of semiclathrate hydrates. These applications include hydrogen storage, cold storage, CO_2 capture, and other gas separation processes. TBAB has been the subject of investigation for its ability to form semi clathrate hydrates under ambient conditions of 1 atm and 12°C. In a study the impact of TBAB on the formation of CO_2 + H_2 + TBAB hydrates was investigated [42]. Their results indicated that a solution containing 0.29 mol% TBAB significantly reduced the formation pressure of semi clathrate hydrates from 11.01 to 0.40 MPa at a temperature of 277.85 K. The semiclathrate hydrate structure involves water molecules forming bonds with bromide anions, resulting in water-bromide host structures. Inside these cages, cations act as guests [24, 76]. Studies have demonstrated that even a small concentration of TBAB, as low as 0.29 mol%, can have a substantial impact on the hydrate formation pressure. The presence of TBAB in the system considerably decreases the formation pressure of CO_2 hydrates, enabling hydrate formation at lower pressures and temperatures. [18] conducted a study to evaluate the CO_2 capture efficiency from a CO_2/H_2 gas mixture (40.2:59.8) using semiclathrate hydrates. The research also examined the kinetics of gas uptake at 3.8 MPa and 273.5 K with 5 wt and 10 wt% (approximately 0.28 mol and 0.56 mol% respectively) TBAB solutions as promoters. The findings indicated that the 10 wt% TBAB solution experiment resulted in higher gas uptake, separation factor, and CO_2 recovery compared to the 5 wt% TBAB solution experiments. A recent research study systematically investigated the impact of different TBAB concentrations on gas absorption and separation effectiveness under conditions of 6.0 MPa and 279.2 K. The TBAB concentrations examined in this study ranged from 0.3 to 3.0 mol%. Notably, a 0.3 mol% TBAB solution exhibited an extended induction time, accompanied by higher overall gas uptake and separation factor. Conversely, when utilizing 1.0 mol% TBAB, a shorter induction time was observed alongside a heightened rate of hydrate growth. Concentrations of TBAB at 1.5, 2.0, and 3.0 mol% exhibited shorter induction times; however, they were associated with decreased gas uptake, sluggish hydrate growth rates, and diminished separation factors [3, 55, 56]. In the research conducted by it was noted that the introduction of TBAB had a substantial impact on the phase boundaries of CO_2 hydrates. When TBAB was present, CO_2 hydrates were observed to develop at lower temperature ranges, spanning from 273.15 K to 291.15 K, and at reduced pressures, ranging from 0.25 MPa to 4.09 MPa. These pressure values were notably lower than those required for pure CO_2 hydrate formation, particularly at lower temperatures [60].

Cyclopentane (CP): The well-known CO_2 gas hydrate promoter cyclopentane (CP) has drawn a lot of interest due to its uses in carbon capture and gas separation. Cyclopentane (CP) is classified as a cycloalkane and exhibits limited miscibility with water. When subjected to standard temperature and pressure conditions, the binary combination of {water + cyclopentane} demonstrates a phenomenon of liquid-liquid phase separation. This results in the formation of distinct liquid phases, one enriched with cyclopentane and the other with water, spanning a broad spectrum of compositions. By lowering the equilibrium pressure and induction time necessary for hydrate formation when added to the gas-water system, CP speeds up the development of CO_2 gas hydrates. Research findings indicate that CP forms sII hydrates exclusively within the spacious water cages, particularly at temperatures close to 280 K and under atmospheric pressure.

Propane: At standard pressures and temperatures, propane functions as a hydrocarbon gas. Its dimensions align well with the structure II hydrates. Incorporating propane into the gas mixture at concentrations ranging from 2.5 to 3.2 mol% leads to a marked reduction in the equilibrium pressure. The evaluation of the heat of dissociation using the Clausius–Clapeyron equation has provided evidence that propane does, in fact, adopt the structure II configuration. The confirmation of structure II formation and the occupancy within its cages has been achieved through a variety of analytical methods, including Raman spectroscopy, infrared analysis, Nuclear Magnetic Resonance (NMR), and Powder X-Ray Diffraction (PXRD). Similar to cyclopentane, propane (C_3H_8) also serves as a promoter for hydrate formation under lowered equilibrium pressure conditions. Propane, when introduced individually, gives rise to structure II (SII) hydrates at a temperature of 275 K and within a pressure range spanning from 0.36 MPa to 0.48 MPa. The existence of propane causes the formation of SII hydrates by competing with CO_2 for occupancy within the larger molecular cages present in the hydrate structure [23, 33, 34, 66]. A study has demonstrated that with the addition of 2.5 mol% C_3H_8 to the system, approximately 57% of the larger cages are occupied by CO_2 [32–34]. In a two-stage clathrate hydrate/membrane process, propane served as a promoter, and it was utilized at a concentration of 2.5 mol%. In this configuration, the first and second stages of the process were conducted at a temperature of 273.7 K, with pressures of 3.8 MPa and 3.5 MPa, respectively. The first stage demonstrated a separation factor of 27.84, a split fraction of 0.47, while the second stage exhibited a separation factor of 91.19 and a split fraction of 0.32. Notably, the addition of propane led to a reduction in the rate of hydrate formation; however, this adjustment did not compromise the efficiency of separation [3–5, 35]. Comprehensive microscopic examinations were carried out to study hydrate formation from gas mixtures containing $CO_2/H_2/C_3H_8$ (with concentrations of 38.1/59.4/2.5 mol%). The objective was to gain a deeper understanding of the mechanism responsible for the increased rate of hydrate formation when propane gas was present within a silica sand bed. The separation factor, which indicates the selectivity of the hydrate formation process for CO_2 over other components, experienced a slight reduction in its value due to the addition of C_3H_8. This suggests that the competitive interaction between CO_2 and C_3H_8 for hydrate cage occupancy, while not drastically altering CO_2 recovery, did lead to a modest decrease in the preference for CO_2 during the hydrate formation process. This decrease in the separation factor could be attributed to the competition between CO_2 and C_3H_8 for available cage sites within the hydrate structure [14].

2.3 Kinetic Hydrate Promoters

Unlike THPs, kinetic hydrate promoter molecules do not occupy the internal space within the hydrate cages, and thus, they do not integrate into the hydrate structure. Consequently, when Kinetic Hydrate Promoters (KHPs) are introduced into the system, they do not bring about a significant change in the hydrate equilibrium curve. Instead, KHPs function by enhancing the initiation of hydrate formation (reducing the time it takes to begin) and expediting the overall process of hydrate development, including its growth rates. In view of environmental considerations, amino acids have emerged as promising candidates for highly effective Kinetic Hydrate Promoters (KHPs). The incorporation of these KHPs can lead to either a reduction in the requisite pressure levels or an elevation in the temperature range wherein Gas Hydrates (GHs) remain stable. Consequently, this adjustment minimizes the energy demand for the compression or cooling of the specific systems in focus [7, 11, 12, 51, 57].

2.3.1 *Surfactants*

Surfactants, recognized as surface-active agents, exhibit remarkable adaptability as chemical compounds that hold a central function in the manipulation of interfaces among distinct phases—liquid, solid, and gas. Their distinctive characteristics empower them to diminish surface tension and foster connections between substances that inherently resist blending. This competence to alter interfacial attributes has propelled their extensive utilization spanning various sectors, encompassing detergents, cosmetics, pharmaceuticals, and agriculture. Within the surfactants' molecular makeup, hydrophilic (water-attraction) and hydrophobic (water-repulsion) segments coexist. This amphiphilic property empowers them to reside at the crossroads of interfaces, where their hydrophilic section engages with water molecules, while their hydrophobic component aligns itself distantly from the aqueous realm. This configuration gives rise to formations such as micelles, monolayers, and bilayers, pivotal in a multitude of processes [20, 27, 31]. In their experimental study [25] examined the impact of surfactants on the kinetics of methane hydrate. Their findings indicated that while surfactants do not alter the thermodynamics, they significantly affect the kinetics of gas dissolution within the water phase. Furthermore, these surfactants exert a substantial influence on the comprehensive rate of hydrate formation. The research conducted by [32] delved into investigating three distinct categories of surfactants—cationic, anionic, and non-ionic—concerning their impact on the kinetics of CO_2 hydrate formation. Notably, among these, the anionic surfactant SDS exhibited the highest efficacy in augmenting the pace of hydrate formation, concurrently curtailing the induction time. Furthermore, the non-ionic surfactant Tween-80 outperformed the cationic surfactant DTACl in the studied context. According to reports, heightened concentrations of SDS have been correlated with increased CO_2 recovery [64]. However, an excessive concentration of SDS could trigger the formation of SDS micelles, thereby resulting in a reduction of contact surfaces between gas and liquid phases and subsequently leading to a decrease in CO_2 recovery. In the study conducted by [27], findings concerning non-ionic surfactants revealed intriguing trends. When present in low concentrations (< 1 wt%), these surfactants appeared to hinder hydrate formation rates compared to systems devoid of any surfactants. Nevertheless, at a concentration of 1 wt% , there was a slight enhancement in hydrate formation rates, approximately amounting to a 20% increase. This implies that non-ionic surfactants typically tend to act more as inhibitors of kinetic hydrate formation rather than promoters. On the other hand, outcomes concerning anionic and cationic surfactants unveiled distinct patterns. These types of surfactants exhibited an optimal concentration range at which they positively influenced the kinetics of hydrate formation. It is worth noting that this optimal concentration is system-specific, contingent upon the hydrate-forming substance and the subcooling levels within the system. The research indicated that the ideal concentration for these surfactants was around 0.05 wt%. Gemini surfactants represent an innovative and distinct category within the surfactant realm. Recognized as dimeric surfactants, these compounds possess two hydrophilic head groups and two hydrophobic tails. What sets these surfactants apart is their distinctive attribute: the hydrophilic heads are linked together through a spacer group of varying lengths. Notably, Gemini surfactants exhibit both reduced Critical Micelle Concentration (CMC) values and decreased surface tension when compared to conventional surfactants, all while functioning in accordance with their respective CMC values [10, 78, 82].

Nanomaterials: Nanomaterials encompass a fascinating class of materials characterized by their dimensions, typically spanning from 1 to 100 nanometres. Their minute size imbues nanomaterials with extraordinary and amplified properties in comparison to their larger counterparts. This unique attribute stems from their capacity to be meticulously engineered and controlled at the atomic and molecular levels, a trait that opens up a vast spectrum of applications across diverse domains. These materials, which include nanoparticles, nanotubes, nanowires, and more, can vary in composition, embracing metals, semiconductors, ceramics, polymers, and even biological molecules [77, 83]. In the year

2006, a pioneering effort was led by [39] as they incorporated copper (Cu) nanoparticles into the gas hydrate formation system, subsequently unveiling their role as promoters in the hydrate formation process. The integration of Cu nanoparticles was shown to effectively enhance the creation of gas hydrates. This phenomenon finds its roots in the remarkable thermal conductivity possessed by nanoparticles, which, upon their incorporation into the solution, led to a substantial enhancement in the overall thermal conductivity of the solution. This revelation underscores the potential of nanoparticles, not only in driving hydrate formation, but also in augmenting the thermal properties of the medium, signifying a significant advancement with implications across diverse applications. In a significant contribution, illuminated [53] the promoting influence exerted by ZnO nanoparticles on the CO_2 hydrate formation. Their findings underscored that the inclusion of ZnO nanoparticles yielded an impressive increase in gas consumption of up to 16.0% when combined with SDS. Broadening their investigation, they ventured into studying the effects of Ag nanoparticles, an SDS solution, and a combined system involving silver (Ag) + SDS on the formation of CO_2 hydrates. Remarkably, the investigation unveiled that the induction time remained relatively consistent across the three scenarios. However, it was exclusively the Ag + SDS combined system that demonstrated an effect on the CO_2 hydrate formation process. This particular combination showcased the ability to substantially elevate gas consumption by 93.3% and amplify the apparent rate constant by an impressive 133.0%. This comprehensive study underscores the intricate interplay between nanoparticles and surfactants, revealing their capacity to significantly modulate the formation dynamics of CO_2 hydrate. An extensive examination encompassed graphene (GP), Graphene Oxide (GO), and Sulphonated Graphene Oxide (SGO) nanosheets in a comparative manner. The outcomes illuminated SGO as the most proficient catalyst across various metrics including the time required for complete hydrate formation, the pace of hydrate formation, and the storage capacity. GP ranked next in terms of efficiency, followed by GO. Notably, GO exhibited subpar promotional effects, a phenomenon attributed to the presence of oxygen-containing groups that have the potential to disrupt the conjugated structure inherent to the nanosheets. This disruption consequently diminished the efficacy of heat transfer, thus influencing the overall hydrate formation process [74]. In a study by [2], they conducted research using hydrophobic silica nanoparticles with an average particle size of 7 nm. The objective of their investigation was to examine the rheological properties of hydrate slurries when subjected to the presence of both a surfactant (Span 80) and silica nanoparticles. At lower nanoparticle concentrations (ranging from 0.05 to 0.5 wt%), these nanoparticles exhibited a behaviour akin to hydrate inhibitors. This was evident through the observed delay in hydrate nucleation, discerned from the viscosity profile data. Intriguingly, as the nanoparticle concentrations escalated to higher levels (≥ 1.0 wt%), an altered scenario emerged. Within this domain, there was a noticeable reduction in the induction time, and hydrate growth proceeded at a faster rate, as evidenced by alterations in the viscosity profile [41, 73].

2.3.2 *Amino Acids*

The distinctive molecular configuration of amino acids underscores their multifaceted roles within biological and chemical frameworks. Structurally, an amino acid is composed of three core constituents: an amino group (-NH_2), a carboxyl group (-COOH), and an R-group, commonly referred to as a side chain. This R-group endows each amino acid with an individual identity, conferring upon it distinctive chemical attributes and behaviours.

2.4 Conclusion

Overall this chapter gives an understanding of the various promoters which are being used to improve the performance of the hydrate formation process. It also covers the interplay between nanoparticles & surfactants, revealing their capacity to moderate the hydrate formation.

References

[1] Adeyemo, A., Kumar, R., Linga, P., Ripmeester, J. and Englezos, P. 2010. Capture of carbon dioxide from flue or fuel gas mixtures by clathrate crystallization in a silica gel column. International Journal of Greenhouse Gas Control 4(3): 478–485. https://doi.org/10.1016/J.IJGGC.2009.11.011.

[2] Ahuja, A., Iqbal, A., Iqbal, M., Lee, J. W. and Morris, J. F. 2018. Rheology of hydrate-forming emulsions stabilized by surfactant and hydrophobic silica nanoparticles. Energy & Fuels 32(5): 5877–5884. https://doi.org/10.1021/acs.energyfuels.8b00795.

[3] Babu, P., Chin, W. I., Kumar, R. and Linga, P. 2014. Systematic evaluation of tetra-n-butyl ammonium bromide (TBAB) for carbon dioxide capture employing the clathrate process. Industrial and Engineering Chemistry Research 53(12): 4878–4887. https://doi.org/10.1021/IE4043714.

[4] Babu, P., Kumar, R. and Linga, P. 2014. Unusual behavior of propane as a co-guest during hydrate formation in silica sand: Potential application to seawater desalination and carbon dioxide capture. Chemical Engineering Science 117: 342–351. https://doi.org/10.1016/J.CES.2014.06.044.

[5] Babu, P., Yang, T., Veluswamy, H. P., Kumar, R. and Linga, P. 2013. Hydrate phase equilibrium of ternary gas mixtures containing carbon dioxide, hydrogen and propane. Journal of Chemical Thermodynamics 61: 58–63. https://doi.org/10.1016/J.JCT.2013.02.003.

[6] Bavoh, C., Lal, B., Osei, H., … K. S.-J. of N. G. and 2019, undefined. n.d. A Review On The Role of Amino Acids in Gas Hydrate Inhibition, CO_2 Capture and Sequestration,

and Natural Gas Storage. Elsevier. Retrieved 19 September 2023, from https://www.sciencedirect.com/science/article/pii/S1875510019300289.

[7] Bhattacharjee, G. and Linga, P. 2021a. Amino acids as kinetic promoters for gas hydrate applications: A mini review. Energy and Fuels 35(9): 7553–7571. https://doi.org/10.1021/ACS.ENERGYFUELS.1C00502.

[8] Bhattacharjee, G. and Linga, P. 2021b. Amino acids as kinetic promoters for gas hydrate applications: A mini review. Energy and Fuels 35(9): 7553–7571. https://doi.org/10.1021/ACS.ENERGYFUELS.1C00502.

[9] Bhattacharjee, G. and Linga, P. 2021c. Amino Acids as Kinetic Promoters for Gas Hydrate Applications: A Mini Review. Energy & Fuels 35(9): 7553–7571. https://doi.org/10.1021/acs.energyfuels.1c00502.

[10] Cai, J., Lv, T., Li, X., Sen, Xu, C. G., von Solms, N. and Liang, X. 2021. Microscopic insights into the effect of the initial gas–liquid interface on hydrate formation by *in-situ* raman in the system of coalbed methane and tetrahydrofuran. ACS Omega 6(51): 35467. https://doi.org/10.1021/ACSOMEGA.1C04907.

[11] Chaturvedi, E., Laik, S., Engineering, A. M.-C. J. of C. and 2021, undefined. n.d. A Comprehensive Review of the Effect of Different Kinetic Promoters on Methane Hydrate Formation. Elsevier. Retrieved 19 September 2023, from https://www.sciencedirect.com/science/article/pii/S1004954120305322.

[12] Cheng, L., Liao, K., Li, Z., Cui, J., Liu, B., Li, F. et al. n.d. The Invalidation Mechanism of Kinetic Hydrate Inhibitors Under High Subcooling Conditions. Elsevier. Retrieved 19 September 2023, from https://www.sciencedirect.com/science/article/pii/S0009250919305354.

[13] Daraboina, N., Engineering, N. von S.-J. of C. and 2015, undefined. n.d. The combined effect of thermodynamic promoters tetrahydrofuran and cyclopentane on the kinetics of flue gas hydrate formation. ACS Publications. Retrieved 19 September 2023, from https://pubs.acs.org/doi/abs/10.1021/je500529w.

[14] Daraboina, N., Ripmeester, J. and Englezos, P. 2013. The impact of SO2 on post combustion carbon dioxide capture in bed of silica sand through hydrate formation. International Journal of Greenhouse Gas Control 15: 97–103. https://doi.org/10.1016/J.IJGGC.2013.02.008.

[15] Delahaye, A., Fournaison, L., Marinhas, S., Chatti, I., Petitet, J. P., Dalmazzone, D. et al. 2006. Effect of THF on equilibrium pressure and dissociation enthalpy of CO_2 hydrates applied to secondary refrigeration. Industrial and Engineering Chemistry Research 45(1): 391–397. https://doi.org/10.1021/IE050356P.

[16] Farrelly, D. J., Everard, C. D., Fagan, C. C. and McDonnell, K. P. 2013. Carbon sequestration and the role of biological carbon mitigation: A review. Renewable and Sustainable Energy Reviews 21: 712–727. https://doi.org/10.1016/J.RSER.2012.12.038.

[17] Gainullin, S. E., Farhadian, A., Kazakova, P. Y., Semenov, M. E., Chirkova, Y. F., Heydari, A. et al. 2023. Novel amino acid derivatives for efficient methane solidification storage via clathrate hydrates without foam formation. Energy and Fuels 37(4): 3208–3217. https://doi.org/10.1021/ACS.ENERGYFUELS.2C03923.

[18] Gholinezhad, J., Chapoy, A. and Tohidi, B. 2011. Thermodynamic stability and self-preservation properties of semi-clathrates in the methane+tetra-n-butyl ammonium bromide+water system. https://researchportal.port.ac.uk/en/publications/thermodynamic-stability-and-self-preservation-properties-of-semi-.

[19] Hashimoto, S., Sugahara, T., Sato, H. and Ohgaki, K. 2007. Thermodynamic stability of H_2 + tetrahydrofuran mixed gas hydrate in nonstoichiometric aqueous solutions. Journal of Chemical and Engineering Data 52(2): 517–520. https://doi.org/10.1021/JE060436F.

[20] He, Y., Sun, M., Chen, C., Zhang, G., ... K. C.-J. of M. and 2019, undefined. n.d. Surfactant-based promotion to gas hydrate formation for energy storage. Pubs.Rsc.Org.

Retrieved 19 September 2023, from https://pubs.rsc.org/en/content/articlehtml/2019/ta/c9ta07071k.
[21] Herslund, P. J., Thomsen, K., Abildskov, J., von Solms, N., Galfré, A., Brântuas, P. et al. 2013. Thermodynamic promotion of carbon dioxide-clathrate hydrate formation by tetrahydrofuran, cyclopentane and their mixtures. IJGGC 17: 397–410. https://doi.org/10.1016/J.IJGGC.2013.05.022.
[22] Herslund, P., Thomsen, K., Equilibria, J. A.-F. P. and 2014, undefined. n.d. Modelling of Cyclopentane Promoted Gas Hydrate Systems for Carbon Dioxide Capture Processes. Elsevier. Retrieved 19 September 2023, from https://www.sciencedirect.com/science/article/pii/S0378381214002593.
[23] Heuvel, M. M.-V. den, … C. P.-F. phase and 2002, undefined. n.d. Gas Hydrate Phase Equilibria for Propane in the Presence of Additive Components. Elsevier. Retrieved 19 September 2023, from https://www.sciencedirect.com/science/article/pii/S0378381201007579.
[24] Jeffrey, G. A. and McMullan, R. K. 2007. The Clathrate Hydrates. 43–108. https://doi.org/10.1002/9780470166093.CH2.
[25] Kalogerakis, N., Jamaluddin, A. K. M., Dholabhai, P. D. and Bishnoi, P. R. 1993. Effect of surfactants on hydrate formation kinetics. Proceedings of the 1993 SPE International Symposium on Oilfield Chemistry, 375–383. https://doi.org/10.2118/25188-MS.
[26] Kang, S. P. and Lee, H. 2000. Recovery of CO_2 from flue gas using gas hydrate: Thermodynamic verification through phase equilibrium measurements. Environmental Science and Technology 34(20): 4397–4400. https://doi.org/10.1021/ES001148L.
[27] Karaaslan, U. and Parlaktuna, M. 2000. Surfactants as hydrate promoters? Energy and Fuels 14(5): 1103–1107. https://doi.org/10.1021/EF000069S.
[28] Kim, S., Choi, S. D. and Seo, Y. 2017. CO_2 capture from flue gas using clathrate formation in the presence of thermodynamic promoters. Energy 118: 950–956. https://doi.org/10.1016/J.ENERGY.2016.10.122.
[29] Kim, S. M., Lee, J. D., Lee, H. J., Lee, E. K. and Kim, Y. 2011. Gas hydrate formation method to capture the carbon dioxide for pre-combustion process in IGCC plant. International Journal of Hydrogen Energy 36(1): 1115–1121. https://doi.org/10.1016/J.IJHYDENE.2010.09.062.
[30] Kumar, A., Bhattacharjee, G., … B. K.-I. and 2015, undefined. n.d. Role of surfactants in promoting gas hydrate formation. ACS Publications. Retrieved 19 September 2023, from https://pubs.acs.org/doi/abs/10.1021/acs.iecr.5b03476.
[31] Kumar, A., Bhattacharjee, G., Kulkarni, B. D. and Kumar, R. 2015a. Role of surfactants in promoting gas hydrate formation. Industrial and Engineering Chemistry Research 54(49): 12217–12232. https://doi.org/10.1021/ACS.IECR.5B03476.
[32] Kumar, A., Bhattacharjee, G., Kulkarni, B. D. and Kumar, R. 2015b. Role of surfactants in promoting gas hydrate formation. Industrial and Engineering Chemistry Research 54(49): 12217–12232. https://doi.org/10.1021/ACS.IECR.5B03476/ASSET/IMAGES/LARGE/IE-2015-034766_0014.JPEG.
[33] Kumar, R., Linga, P., Moudrakovski, I., Ripmeester, J. A. and Englezos, P. 2008a. Structure and kinetics of gas hydrates from methane/ethane/propane mixtures relevant to the design of natural gas hydrate storage and transport facilities. AIChE Journal 54(8): 2132–2144. https://doi.org/10.1002/AIC.11527.
[34] Kumar, R., Linga, P., Moudrakovski, I., Ripmeester, J. A. and Englezos, P. 2008b. Structure and kinetics of gas hydrates from methane/ethane/propane mixtures relevant to the design of natural gas hydrate storage and transport facilities. AIChE Journal 54(8): 2132–2144. https://doi.org/10.1002/AIC.11527.
[35] Kumar, R., Linga, P., Ripmeester, J. A. and Englezos, P. 2009. Two-stage clathrate hydrate/membrane process for precombustion capture of carbon dioxide and hydrogen.

Journal of Environmental Engineering 135(6): 411–417. https://doi.org/10.1061/(ASCE)EE.1943-7870.0000002.
[36] Lee, H. J., Lee, J. D., Linga, P., Englezos, P., Kim, Y. S., Lee, M. S. et al. 2010. Gas hydrate formation process for pre-combustion capture of carbon dioxide. Energy 35(6): 2729–2733. https://doi.org/10.1016/J.ENERGY.2009.05.026.
[37] Lee, W., Kim, Y., journal, S. K.-C. engineering and 2018, undefined. n.d. Semiclathrate-based CO_2 Capture from Fuel Gas in the Presence of Tetra-N-Butyl Ammonium Bromide and Silica Gel Pore Structure. Elsevier. Retrieved 19 September 2023, from https://www.sciencedirect.com/science/article/pii/S1385894717314493.
[38] Lee, Y. J., Kawamura, T., Yamamoto, Y. and Yoon, J. H. 2012. Phase equilibrium studies of tetrahydrofuran (THF) + CH_4, THF + CO_2, CH_4 + CO_2, and THF + CO_2 + CH_4 hydrates. Journal of Chemical and Engineering Data 57(12): 3543–3548. https://doi.org/10.1021/JE300850Q/ASSET/IMAGES/LARGE/JE-2012-00850Q_0006.JPEG.
[39] Li, J., Liang, D., Guo, K., Wang, R. and Fan, S. 2006a. Formation and dissociation of HFC134a gas hydrate in nano-copper suspension. Energy Conversion and Management 47(2): 201–210. https://doi.org/10.1016/j.enconman.2005.03.018.
[40] Li, J., Liang, D., Guo, K., Wang, R. and Fan, S. 2006b. Formation and dissociation of HFC134a gas hydrate in nano-copper suspension. Energy Conversion and Management 47(2): 201–210. https://doi.org/10.1016/j.enconman.2005.03.018.
[41] Li, S., Fan, S., Wang, J., Lang, X., Chemistry, D. L.-J. of N. G. and 2009, undefined. n.d. CO_2 Capture from Binary Mixture Via Forming Hydrate with the Help of Tetra-N-Butyl Ammonium Bromide. Elsevier. Retrieved 19 September 2023, from https://www.sciencedirect.com/science/article/pii/S1003995308600857.
[42] Li, X. Sen, Xia, Z. M., Chen, Z. Y., Yan, K. F., Li, G. and Wu, H. J. 2010. Equilibrium hydrate formation conditions for the mixtures of CO_2 + H_2 + tetrabutyl ammonium bromide. Journal of Chemical and Engineering Data 55(6): 2180–2184. https://doi.org/10.1021/JE900758T/ASSET/IMAGES/LARGE/JE-2009-00758T_0001.JPEG.
[43] Li, X. Sen, Xu, C. G., Chen, Z. Y. and Wu, H. J. 2011. Hydrate-based pre-combustion carbon dioxide capture process in the system with tetra-n-butyl ammonium bromide solution in the presence of cyclopentane. Energy 36(3): 1394–1403. https://doi.org/10.1016/j.energy.2011.01.034.
[44] Linga, P., Kumar, R. and Englezos, P. 2007a. Gas hydrate formation from hydrogen/carbon dioxide and nitrogen/carbon dioxide gas mixtures. Chemical Engineering Science 62(16): 4268–4276. https://doi.org/10.1016/J.CES.2007.04.033.
[45] Linga, P., Kumar, R. and Englezos, P. 2007b. The clathrate hydrate process for post and pre-combustion capture of carbon dioxide. Journal of Hazardous Materials 149(3): 625–629. https://doi.org/10.1016/J.JHAZMAT.2007.06.086.
[46] Liu, N., Huang, J., Meng, F., L. Y.-A. S. C. and 2023, undefined. n.d. Experimental study on the mechanism of enhanced CO_2 hydrate generation by thermodynamic promoters. ACS Publications. Retrieved 19 September 2023, from https://pubs.acs.org/doi/abs/10.1021/acssuschemeng.2c05655.
[47] Liu, Y., Chen, B., Chen, Y., Zhang, S., … W. G.-E. and 2015, undefined. n.d. Methane Storage in a Hydrated form as Promoted by Leucines for Possible Application to Natural Gas Transportation and Storage. Wiley Online Library. Retrieved 19 September 2023, from https://onlinelibrary.wiley.com/doi/abs/10.1002/ente.201500048.
[48] Liu, Y., Chen, B., Chen, Y., Zhang, S., Guo, W., Cai, Y. et al. 2015a. Methane storage in a hydrated form as promoted by leucines for possible application to natural gas transportation and storage. Energy Technology 3(8): 815–819. https://doi.org/10.1002/ente.201500048.
[49] Liu, Y., Chen, B., Chen, Y., Zhang, S., Guo, W., Cai, Y. et al. 2015b. Methane storage in a hydrated form as promoted by leucines for possible application to natural gas

transportation and storage. Energy Technology 3(8): 815–819. https://doi.org/10.1002/ENTE.201500048.

[50] Majid, A. A. A., Worley, J. and Koh, C. A. 2021a. Thermodynamic and kinetic promoters for gas hydrate technological applications. Energy and Fuels 35(23): 19288–19301. https://doi.org/10.1021/ACS.ENERGYFUELS.1C02786/ASSET/IMAGES/LARGE/EF1C02786_0017.JPEG.

[51] Majid, A. A. A., Worley, J. and Koh, C. A. 2021b. Thermodynamic and kinetic promoters for gas hydrate technological applications. Energy and Fuels 35(23): 19288–19301. https://doi.org/10.1021/ACS.ENERGYFUELS.1C02786.

[52] Mohammadi, A., Afzal, W., chemical, D. R.-T. journal of and 2008, undefined. n.d. Gas Hydrates of Methane, Ethane, Propane, and Carbon Dioxide in the Presence of Single NaCl, KCl, and $CaCl_2$ Aqueous Solutions: Experimental. Elsevier. Retrieved 19 September 2023, from https://www.sciencedirect.com/science/article/pii/S0021961408001432.

[53] Mohammadi, M., Haghtalab, A. and Fakhroueian, Z. 2016. Experimental study and thermodynamic modeling of CO_2 gas hydrate formation in presence of zinc oxide nanoparticles. The Journal of Chemical Thermodynamics 96: 24–33. https://doi.org/10.1016/j.jct.2015.12.015.

[54] Nanda, S., Reddy, S. N., Mitra, S. K. and Kozinski, J. A. 2016. The progressive routes for carbon capture and sequestration. Energy Science and Engineering 4(2): 99–122. https://doi.org/10.1002/ESE3.117.

[55] Park, S., Lee, S., Lee, Y., Lee, Y. and Seo, Y. 2013. Hydrate-based pre-combustion capture of carbon dioxide in the presence of a thermodynamic promoter and porous silica gels. International Journal of Greenhouse Gas Control 14: 193–199. https://doi.org/10.1016/J.IJGGC.2013.01.026.

[56] Park, S., Lee, S., Lee, Y. and Seo, Y. 2013. CO_2 capture from simulated fuel gas mixtures using semiclathrate hydrates formed by quaternary ammonium salts. Environmental Science and Technology 47(13): 7571–7577. https://doi.org/10.1021/ES400966X/SUPPL_FILE/ES400966X_SI_001.PDF.

[57] Partoon, B. and Javanmardi, J. 2013. Effect of mixed thermodynamic and kinetic hydrate promoters on methane hydrate phase boundary and formation kinetics. Journal of Chemical and Engineering Data 58(3): 501–509. https://doi.org/10.1021/JE301153T.

[58] Rossi, F., Gambelli, A. M., Sharma, D. K., Castellani, B., Nicolini, A. and Castaldi, M. J. 2019. Experiments on methane hydrates formation in seabed deposits and gas recovery adopting carbon dioxide replacement strategies. Applied Thermal Engineering 148: 371–381. https://doi.org/10.1016/J.APPLTHERMALENG.2018.11.053.

[59] Seo, Y., Lee, S., Cha, I., Lee, J. D. and Lee, H. 2009. Phase equilibria and thermodynamic modeling of ethane and propane hydrates in porous silica gels. Journal of Physical Chemistry B 113(16): 5487–5492. https://doi.org/10.1021/JP810453T.

[60] Shimada, W., Ebinuma, T., Oyama, H., Kamata, Y., Takeya, S., Uchida, T. et al. 2003. Separation of gas molecule using tetra-n-butyl ammonium bromide semi-clathrate hydrate crystals. Japanese Journal of Applied Physics, Part 2: Letters 42(2 A). https://doi.org/10.1143/JJAP.42.L129.

[61] Stuart Haszeldine, R. 2009. Carbon capture and storage: how green can black be? Science 325(5948): 1647–1652. https://doi.org/10.1126/SCIENCE.1172246.

[62] Sun, Z., Fan, S., Guo, K., Shi, L., ... Y. G.-J. of C. and 2002, undefined. n.d. Gas hydrate phase equilibrium data of cyclohexane and cyclopentane. ACS Publications. Retrieved 19 September 2023, from https://pubs.acs.org/doi/abs/10.1021/je0102199.

[63] Sun, Z. G., Fan, S. S., Guo, K. H., Shi, L., Guo, Y. K. and Wang, R. Z. 2002. Gas hydrate phase equilibrium data of cyclohexane and cyclopentane. Journal of Chemical and Engineering Data 47(2): 313–315. https://doi.org/10.1021/JE0102199.

[64] Tang, J., Zeng, D., Wang, C., Chen, Y., He, L. and Cai, N. 2013. Study on the influence of SDS and THF on hydrate-based gas separation performance. Chemical Engineering Research and Design 91(9): 1777–1782. https://doi.org/10.1016/J.CHERD.2013.03.013.

[65] Tzirakis, F., Stringari, P., Coquelet, C., von Solms, N. and Kontogeorgis, G. 2016. Hydrate equilibrium data for CO_2 + N_2 system in the presence of tetra-n-butylammonium fluoride (TBAF) and mixture of TBAF and Cyclopentane (CP). Journal of Chemical & Engineering Data 61(2): 1007–1011. https://doi.org/10.1021/acs.jced.5b00942.

[66] Uchida, T., Moriwaki, M., Takeya, S., Ikeda, I. Y., Ohmura, R., Nagao, J. et al. 2004. Two-step formation of methane–propane mixed gas hydrates in a batch-type reactor. Wiley Online Library 50(2): 518–523. https://doi.org/10.1002/aic.10045.

[67] Veluswamy, H., Hong, Q., Design, P. L.-C. G. and 2016, undefined. n.d. Morphology Study of Methane Hydrate Formation and Dissociation in the Presence of Amino Acid. ACS Publications. Retrieved 19 September 2023, from https://pubs.acs.org/doi/abs/10.1021/acs.cgd.6b00997.

[68] Veluswamy, H. P., Hong, Q. W. and Linga, P. 2016. Morphology study of methane hydrate formation and dissociation in the presence of amino acid. Crystal Growth & Design 16(10): 5932–5945. https://doi.org/10.1021/acs.cgd.6b00997.

[69] Veluswamy, H. P., Lee, P. Y., Premasinghe, K. and Linga, P. 2017a. Effect of biofriendly amino acids on the kinetics of methane hydrate formation and dissociation. Industrial and Engineering Chemistry Research 56(21): 6145–6154. https://doi.org/10.1021/ACS.IECR.7B00427.

[70] Veluswamy, H. P., Lee, P. Y., Premasinghe, K. and Linga, P. 2017b. Effect of biofriendly amino acids on the kinetics of methane hydrate formation and dissociation. Industrial & Engineering Chemistry Research 56(21): 6145–6154. https://doi.org/10.1021/acs.iecr.7b00427.

[71] Vlasic, T. M., Servio, P. D. and Rey, A. D. 2019a. THF hydrates as model systems for natural gas hydrates: comparing their mechanical and vibrational properties. Industrial and Engineering Chemistry Research 58(36): 16588–16596. https://doi.org/10.1021/ACS.IECR.9B02698.

[72] Vlasic, T. M., Servio, P. D. and Rey, A. D. 2019b. THF hydrates as model systems for natural gas hydrates: comparing their mechanical and vibrational properties. Industrial and Engineering Chemistry Research 58(36): 16588–16596. https://doi.org/10.1021/ACS.IECR.9B02698.

[73] Wang, F., Luo, S.-J., Fu, S.-F., Jia, Z.-Z., Dai, M., Wang, C.-S. et al. 2015. Methane hydrate formation with surfactants fixed on the surface of polystyrene nanospheres. Journal of Materials Chemistry A 3(16): 8316–8323. https://doi.org/10.1039/C5TA01101A.

[74] Wang, F., Meng, H.-L., Guo, G., Luo, S.-J. and Guo, R.-B. 2017. Methane hydrate formation promoted by $-SO_3^-$-coated graphene oxide nanosheets. ACS Sustainable Chemistry & Engineering 5(8): 6597–6604. https://doi.org/10.1021/acssuschemeng.7b00846.

[75] Wang, Y., Lang, X., Fan, S., Wang, S., Yu, C. and Li, G. 2021. Review on enhanced technology of natural gas hydrate recovery by carbon dioxide replacement. Energy and Fuels 35(5): 3659–3674. https://doi.org/10.1021/ACS.ENERGYFUELS.0C04138/ASSET/IMAGES/LARGE/EF0C04138_0009.JPEG.

[76] Ye, N. and Zhang, P. 2012. Equilibrium data and morphology of tetra-n-butyl ammonium bromide semiclathrate hydrate with carbon dioxide. Journal of Chemical and Engineering Data 57(5): 1557–1562. https://doi.org/10.1021/JE3001443.

[77] Yu, Y., Xu, C. and Li, X. 2018. Evaluation of CO_2 hydrate formation from mixture of graphite nanoparticle and sodium dodecyl benzene sulfonate. Journal of Industrial and Engineering Chemistry 59: 64–69. https://doi.org/10.1016/j.jiec.2017.10.007.

[78] Zhang, J. S., Lo, C., Somasundaran, P. and Lee, J. W. 2010. Competitive adsorption between SDS and carbonate on tetrahydrofuran hydrates. Journal of Colloid and Interface Science 341(2): 286–288. https://doi.org/10.1016/j.jcis.2009.09.052.
[79] Zhao, W., Zhong, D., Equilibria, C. Y.-F. P. and 2016, undefined. n.d. Prediction of Phase Equilibrium Conditions for Gas Hydrates Formed in the Presence of Cyclopentane or Cyclohexane. Elsevier. Retrieved 19 September 2023, from https://www.sciencedirect.com/science/article/pii/S037838121630320X.
[80] Zheng, J., Chong, Z. R., Qureshi, M. F. and Linga, P. 2020. Carbon dioxide sequestration via gas hydrates: a potential pathway toward decarbonization. Energy and Fuels 34(9): 10529–10546. https://doi.org/10.1021/ACS.ENERGYFUELS.0C02309/ASSET/IMAGES/LARGE/EF0C02309_0010.JPEG.
[81] Zhong, D. L., Li, Z., Lu, Y. Y. and Sun, D. J. 2014. Phase equilibrium data of gas hydrates formed from a CO_2 + CH_4 gas mixture in the presence of tetrahydrofuran. Journal of Chemical and Engineering Data 59(12): 4110–4117. https://doi.org/10.1021/JE5007482.
[82] Zhong, Y. and Rogers, R. E. 2000. Surfactant effects on gas hydrate formation. Chemical Engineering Science 55(19): 4175–4187. https://doi.org/10.1016/S0009-2509(00)00072-5.
[83] Zhou, S., Jiang, K., Zhao, Y., Chi, Y., Wang, S. and Zhang, G. 2018. Experimental investigation of CO_2 hydrate formation in the water containing graphite nanoparticles and tetra-n-butyl ammonium bromide. Journal of Chemical & Engineering Data 63(2): 389–394. https://doi.org/10.1021/acs.jced.7b00785.

Chapter 3

Feasibility of Long-Distance Transportation of Captured CO_2 using Transmission Pipelines

An Overview of Techno-economic Models

Ali U. Chaudhry,[1,*] *Fadhel T. Abdullah*,[2]
Muhammad Saad Khan,[1] *Harris Sajjad Rabbani*[1] and *Bhajan Lal*[3]

3.1 Introduction

The Paris Climate Accord in December 2015 within United Nations Framework Convention on Climate Change (UNFCCC) aimed "*to hold the increase in the global average temperature to well below 2°C above preindustrial levels and to pursue efforts to limit the temperature increase to 1.5°C above pre-industrial levels*" [1, 2]. Carbon dioxide (CO_2) is one of the main constituents (68% of the total) of anthropogenic greenhouse gases (GHG) and air toxics, and correlated with increase in global average temperature [3]. According to a report, approximately 13,466 $MtCO_2/yr$[2] of manmade CO_2 emissions were recorded in 2005. It was also reported that a significant fraction, i.e., 85% of energy used globally comes from fossil fuels which in turn discharges 30 billion tons of carbon dioxide [4, 5]. It is also estimated that by 2050 CO_2 emissions will be increased by 130%. In this scenario, the grand challenge of a clean environment and stabilization of GHG

[1] Department of Petroleum Engineering, Texas A&M University at Qatar, Qatar.
[2] Department of Mechanical Engineering, College of Engineering, Al Asala Colleges, Saudi Arabia.
[3] Chemical Engineering Department, Center for Carbon Capture, Storage and Utilization (CCUS), Institute of Sustainable Energy, Universiti Teknologi PETRONAS, Bandar Seri Iskandar, Perak, Malaysia.
* Corresponding author: chaudhry.usman@qatar.tamu.edu

concentration may be achieved earnestly with the help of non-fossil energy sources or renewable energy, improved energy efficiency, nuclear energy, and near-decarburization of fossil fuel-based power generation [6, 7]. However, some obstacles hinder the shifting and implementation of current fossil energy technological systems to non-fossil systems. Due to technological transition delay and abundant fossil fuels, the energy obtained from fossil fuels will remain a large contributor in future energy demands for several decades.

To enable the continuing use of fossil fuels while reducing the emissions of CO_2, (CCS) is one of the very few methods available to mitigate the anthropogenic impact on the environment significantly and relatively urgently at a realistic cost. CCS involves the capturing and squeezing of CO_2 from different industrial sites like coal fire plants, ethanol fermentation, cement manufacturers, power generation plants, fertilizers, steel making, etc. The captured CO_2 can be stored securely into depleted oil and gas fields, saline aquifers, and deep-sea sediments (Table 1) for hundreds or thousands of years [8]. Table 2 summarizes major viable sink options on the ocean and land, and their capacities. Based on worldwide total CO_2 emissions (~ 7 GtC per year), these storage options together indicates that a large capacity of CO_2 can be stored especially in the ocean. It can also be seen from Table 2 that compared to the atmosphere and terrestrial biosphere, oceans are a significant sink for CO_2 because they already contain an estimated 40,000 GtC (billion metric tons of carbon) and adding carbon would only change the ocean's concentration by less than 2% [9]. Depending on available storage capacities, CSS may act as an interim system to reduce the CO_2 emission with continuous use of fossil fuels during a 50 year transition [10]. Figure 1 shows the graphical representation of CSS liberated from combustion by biological and non-biological processes. As can be seen from the CSS chain, CO_2 obtained from different combustion processes can be captured biologically using trees or microalgae and transported and stored at different geological locations [11].

Numerous review papers cover specific aspects of CCUS literature, such as CCU impacts, integration with CCS, techno-economic analysis,

Table 1. Estimated world sink capabilities of CO_2 disposal options [9].

Sequestration option	**Worldwide capacity (Orders of magnitude estimates)**
Ocean	1000 GtC
Deep saline formations	100–1000 GtC
Depleted oil and gas reservoirs	100 GtC
Coal seams	10–100 GtC
Terrestrial	10 GtC
Utilization	< 1 GtC/year

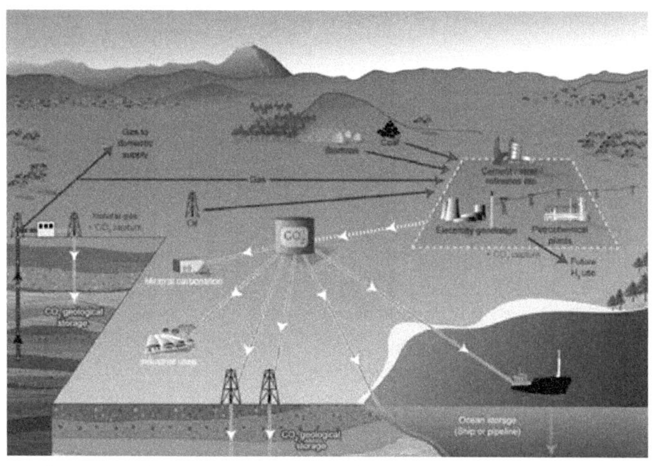

Figure 1. Schematic view of CCS chain from source to sink [12].

and planning. One paper examines CO_2 utilization options, while another surveys the state-of-the-art in LCA of CCUS systems, differentiating between carbon management benefits of CCU and CCS, and explores CO_2 utilization modelling, including Enhanced Oil Recovery while other papers discuss technological advances in capture, utilization, and storage, as well as further explore monitoring and verification techniques.

3.2 Thermodynamic Properties of CO_2 for Pipeline Transportation

CCS systems depend on CO_2 mixture thermodynamics for design and operation. Operational parameters' impact on performance and costs relies on CO_2 mixture properties. For instance, vapour liquid equilibrium is crucial in designing CO_2 purification from coal-fired power plant flue gas. High-density CO_2 transport, avoiding two-phase flow, reduces energy and investment costs and ensures safety. Accurate thermodynamic knowledge is vital to control CCS system parameters for a proper operation. To evaluate CO_2 thermodynamic properties in CCS operations, an operating window must be established by considering the phase regions and CCS processes. Determining data requirements relies on temperature and pressure conditions, guiding the use of property models to minimize errors. CCS operations involve CO_2 capture from flue gas, processing (compression, dehydration, purification/liquefaction, and additional compression/pumping), transport, and storage. These phases form the CCS process chain. Table 2 (adopted form Hailong Li Thesis) estimates pressure and temperature conditions for CCS operations, including subprocesses. Figure 2 illustrates P-T windows, shaped by projected operational conditions in CCS processes.

Table 2. Estimated operation conditions (P and T) of the CCS processes (adopted from Hailong Li Thesis).

CCS process	T (K)	P (MPa)
CO_2 purification and compression	***219.15 to 423.15***	***0 to 11***
Initial compression	293.15 to 423.15	0 to 3
Dehydration	283.15 to 303.15	2 to 3
Purification	219.15 to 248.15	2 to 5
Further compression and pumping	283.15 to 303.15	5 to 11
CO_2 transport	***218.15 to 303.15***	***0.5 to 20***
Pipeline	273.15 to 303.15	7.5 to 20
Small tanks	238.15 to 248.15	1.5 to 2.5
Large tanks	218.15 to 228.15	0.5 to 0.9
CO_2 storage	***277.15 to 423.15***	***0.1 to 50***

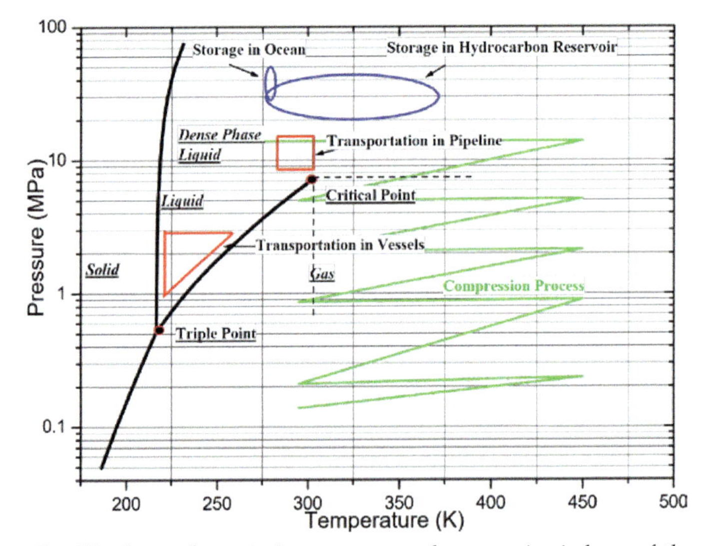

Figure 2. Possible thermodynamic (temperature and pressure) windows of the effective CCS systems (adopted from Hailong Li Thesis).

3.3 Captured Carbon Transportation (CCT)

CCT through pipeline has been recognized as one of several approaches necessary for transporting CO_2 emission from the point of source to the point of storage. As mentioned in the IPCC report published in 2000, 60% of CO_2 emission were related to 7887 sources whereas the remaining 40% were mainly attributed to transportation systems [13]. There is a requirement of committed CO_2 pipeline network to connect the widely and geographical distributed reservoirs or storage (deep saline aquifers, oil and gas wells, and un-mineable coal reservoirs [14]) and CO_2 source of emission [8]. A normal 1000-megawatt natural gas fired power station usually releases about 3 million tons CO_2 which is about one-half of the emission liberated by a pulverized coal-fired power plant. To store these emissions, it would require a building with a base like a football square with height of 750 m [15]. Skovholt studied a feasibility of a transportation unit comprising of a single 1000 MW gas fired power station attached with a large diameter pipeline network over 1500 km long. The study shows that utilizing a large diameter pipeline compare to 16 inch pipeline to transport CO_2 can reduce the cost from USD$ 42 to $6 per ton over the same distance [15]. In perspective, it is also estimated that by 2030, coal will be supplying 28% of the global energy which in turn will increase the CO_2 by 57% [10]. Figure 2 depicts a scheme of a pilot stage hypothetical town attached to a 1000 megawatt coal plant (which emits between 6 and 8 Mt/yr of CO_2) [13] for electricity generation along with a CO_2 sequestering plant. It can be seen from the depiction that a well-connected pipeline system would be required to pump the emitted CO_2 from coal plant to more than two kilometres below the surface into the microscopic pore space of sedimentary rocks spread over 40 square kilometres [16]. However, transportation of emitted CO_2 by road, rail or ship is unthinkable and would result in high unit cost of transport. In the US and Canada, compressed CO_2 emission has already been transported using 3000 km of operational pipeline (Table 1) system since the 1970s which transports around 30 MT CO_2 per year. It can be seen from Table 1 that Shute Creek is the largest with a diameter of 30 inch, 808 km in length and flow capacity of 19.3 million tons/year. According to MIT, if 60% of the CO_2 emitted from the US coal power plant were to be transported in liquid form for geologic storage, its volume would be approximately equal to the total US oil consumption of about 20 million barrels per day. This means that a massive scale infrastructure and detailed planning would be required. The designing, modelling and optimization of such massive scale infrastructure would require to considering key decisions such as the amount of CO_2 to capture, their sources, location of pipelines and reservoirs size of pipelines, amount of CO_2 injection, and distribution [3]. In Europe, the proposed map of shared pipeline network for CO_2 transportation can be seen in Figure 3. This pipeline system can collect compressed CO_2 at 150 bar and be conveyed to spatially suitable offshore nodes or individual storage sites [10, 17, 18]. The most significant development has been seen in the Sleipner Project located in the North Sea about 240 km off the Norwegian cost. At this site, about 1 million metric tons of compressed CO_2 is stored annually into a 200-m-thick sandstone layer lies about 1000 m below the seabed as shown in Table 3. Based on the Norwegian offshore carbon tax,

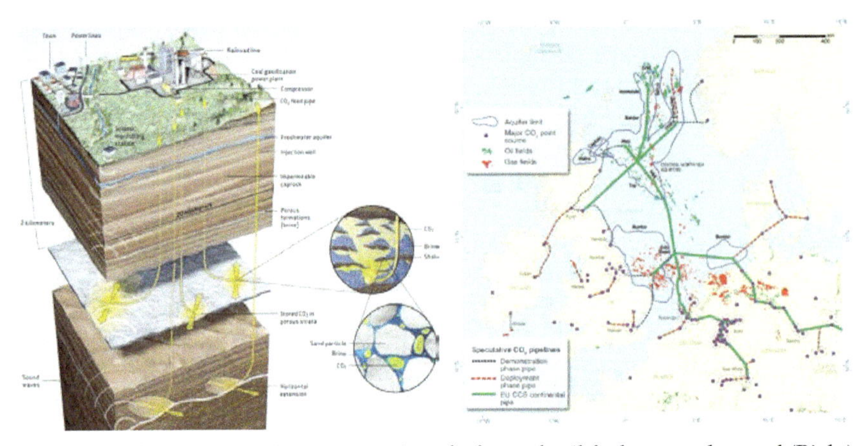

Figure 3. (Left) Diagrammatic representation of a future fossil-fuel power plant and (Right) Map of northwest Europe, showing sites of emissions, saline formations, gas fields, and oil fields. CO_2 can be collected from clusters of large power plants and transported to storage. This transport scenario visualizes pipelines built to offshore hubs accessing large-scale storage beneath the North Sea. Such sites can be evaluated with the use of legacy hydrocarbon data and may prove to be more reliable to develop and monitor than onshore storage [10, 16].

Table 3. Existing U.S. CO_2 pipelines, lengths, mass flow rates and diameters [14].

Pipeline	Location	Pipe length (km)	Mass flow rate (million tons per year)	Pipe diameter (inch)
Bravo	New Mexico/Texas, USA	351	7.4	20″
Canyon reef carriers	Texas, USA	225	4.4	16″
Cortez	Texas, USA	808	19.3	30″
Sheep Mountain-1	Texas, USA	296	6.4	20″
Sheep Mountain-2	Texas, USA	360	9.3	24″
Tranpetco	Texas and Oklahoma, USA	193	3.4	12″
Val verde	Texas, USA	130	2.5	10″
Weyburn	North Dakota, USA and Saskatchewan, Canada	330	1.8	12″

the cost of the project was paid back in about one and half years indicating the successful commercialization of CO_2 sequestration [9].

Among, all the available options for CO_2 transportation, i.e., tanks, trains, trucks and ships, pipelines are the most viable form of long-distance transport. The design and economics of the pipeline network may depend on the following primary factors, i.e., pipeline route selection, design life, the network shape and constructability, and operating philosophy.

Other secondary factors are public and personnel safety, operating and geological conditions, environmental impact, ease of construction, maintenance, security, pipeline diameter and length, operating pressure, and any other hazards [18]. In literature, most of the focus has been placed on developing models for estimating the cost and design of the pipeline carrying dense phase CO_2, while experimental work received relatively less attention. These models explain the economies of scale that can be reached for the collection of CO_2 emissions from multiple sources and transporting them into pipelines to a single storage site [9]. The cost of the CCS transportation mainly depends on the four main factors source of CO_2; transportation, distance, and the type and characteristics of the sequestration reservoir.

Introduction of a scalable infrastructure model for CCS that generates a fully integrated, cost-minimizing CCS system [8]. The model determines where and how much CO_2 to capture and store, and where to build and connect pipelines of different sizes, in order to minimize the combined annualized costs of sequestering a given amount of CO_2. The model was able to aggregate CO_2 flows between sources and reservoirs into trunk pipelines that take advantage of economies of scale. Pipeline construction costs take into account factors including topography and social impacts. The model can be used to calculate the scale of CCS deployment (local, regional, national). The researcher demonstrated that the model can use a set of 37 CO_2 sources and 14 reservoirs for California. The results highlight the importance of systematic planning for CCS infrastructure by examining the sensitivity of CCS infrastructure, as optimized by the model, to varying CO_2 targets [8]. Morbee et al. studied the extent and cost of the optimal least-cost CO_2 transport network at European scale for the period 2015–2050. The study started in 2015, the earliest foreseeable starting date of the CCS projects co-funded by the European Energy Programme for Recovery (EEPR), with 2050 as the EU's target date for 80–95% reduction of greenhouse gas emissions. Their study showed that by 2030, more than half of the EU member states could be involved in cross-border CO_2 transport (Figure 4) [19]. Svensson et al. analyzed the feasibility of pipelines for CO_2 transportation using 1 Mt/y of CO_2 to 300 Mt/yr of CO_2. The transportation costs for the demonstration plant scenario range from 1 to 6 euro/ton of CO_2 depending on storage type and means of transportation. The study suggested that a combination of pipelines and water carriers is the most cost effective alternative for off shore transportation [20].

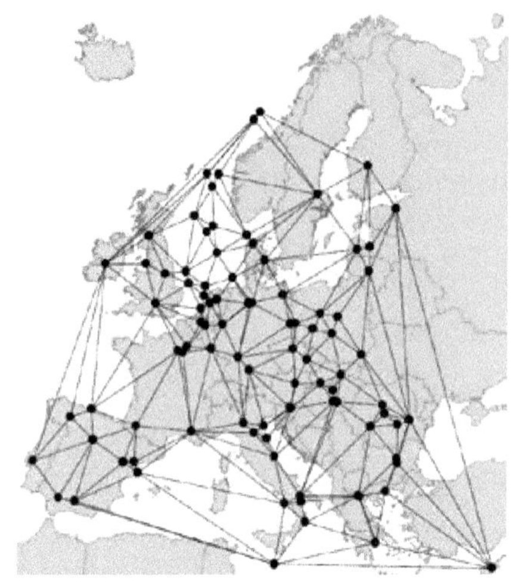

Figure 4. Delaunay triangulation to identify candidate pipelines [19].

3.4 Conclusion

The chapter provides insights into the challenges and opportunities associated with CCS and the transportation of captured carbon. It highlights the significance of addressing CO_2 emissions from various sources through well-designed infrastructure and economic models, contributing to the global efforts in mitigating climate change. The precise understanding of the thermodynamics of CO_2 is imperative for effective control of CCS system parameters, ensuring optimal operation. The establishment of an operating window, taking into account phase regions and CCS processes, is essential to evaluate CO_2 thermodynamic properties. The determination data requirements hinges on temperature and pressure conditions, necessitating the use of property models to minimize errors. This emphasis on accurate thermodynamic knowledge underscores the foundational role it plays in the successful implementation and performance of CCS operations. Pipelines emerge as the most feasible means for long-distance CO_2 transportation when compared to alternative options such as tanks, trains, trucks, and ships. The economic and design aspects of the pipeline network are contingent upon primary factors like pipeline route selection, design life, network configuration, and constructability, as well as secondary considerations including public and personnel safety, operating and geological conditions, environmental impact, ease of construction and maintenance, security, pipeline diameter and length, operating pressure, and other potential hazards. The research indicates that opting for a larger diameter pipeline can result in a substantial cost reduction from over the same distance. The strategic planning,

Table 4. Overview of the economic models in literature [22].

No.	Type of model	Original goal of the cost model	Location	Pressure Inlet (MPa)	Region specific	Costs include
1	Linear [23]	Development of a large-scale CO_2 infrastructure till 2050	Onshore and offshore	11	Netherlands	Material, construction, insurance, licenses, and engineering costs
2	Linear [24]	Assessing the economics of CO_2 transport and storage.	Onshore and offshore	15.2	U.S.	Material, labor, ROW, and misc. costs
3	Linear [25]	Optimize worldwide source and sink connections for 2030 and 2050.	Onshore and offshore	15.0 (13–20)	World oriented with regional factor	Material, labor, ROW, contingencies, and owners' cost
4	Weight based model [26]	Estimate cost for different modes of CO_2 transport (ships, pipeline, and train) for a given case study	Onshore	15.2	China	Material, labor, ROW and misc. costs
5	Weight based model [27]	Building a projection tool for CCS in Belgium till 2050	Onshore	12.5	Regional factor included in material costs (for Belgium).	Material, labor, ROW, and misc. costs

modelling, and optimization of such extensive infrastructure demand careful consideration of key decisions, including the volume of CO_2 to capture, their sources, pipeline and reservoir locations, pipeline size, CO_2 injection amounts, and distribution logistics. Furthermore, the studies propose that the combination of pipelines and water carriers stands out as the most economically viable choice for offshore transportation. The economic models which have been used for costing in the literature have been mentioned in Table 4.

References

[1] Agreement, P. 2015. Adoption of the Paris Agreement. FCCC/CP/2015/L. 9/Rev. 1.

[2] Joeri Rogelj, Michel den Elzen, Niklas Höhne, Taryn Fransen, Hanna Fekete and Harald Winkler. 2016. Paris Agreement climate proposals need a boost to keep warming well below 2°C. Nature 534: 631.

[3] Deutch, J. and Moniz, E. J. 2007. The Future of Coal: Options for a Carbon-Constrained World. Cambridge: Massachusetts Institute of Technology. 63.

[4] Niall MacDowell, Nick Florin, Antoine Buchard, Jason Hallett, Amparo Galindo, George Jackson et al. 2010. An overview of CO_2 capture technologies. Energy & Environmental Science 3(11): 1645–1669.

[5] Thompson, L. 2011. Applications: Energy from fossil resources. pp. 151–183. *In*: International Assessment of Research and Development in Catalysis by Nanostructured Materials. Imperial College Press.

[6] Chad A. Mirkin and Mark C. Hersam. 2011. Applications: Catalysis by nanostructured materials. pp. 445–466. *In*: Nanotechnology Research Directions for Societal Needs in 2020: Retrospective and Outlook. Springer Netherlands: Dordrecht.

[7] Oh, T. H. 2010. Carbon capture and storage potential in coal-fired plant in Malaysia—A review. Renewable and Sustainable Energy Reviews 14(9): 2697–2709.

[8] Middleton, R. S. and Bielicki, J. M. 2009. A scalable infrastructure model for carbon capture and storage: SimCCS. Energy Policy 37(3): 1052–1060.

[9] Herzog, H. J. 2001. Peer Reviewed: What Future for Carbon Capture and Sequestration? ACS Publications.

[10] Haszeldine, R. S. 2009. Carbon capture and storage: how green can black be? Science 325(5948): 1647–1652.

[11] Stewart, C. and Hessami, M.-A. 2005. A study of methods of carbon dioxide capture and sequestration—the sustainability of a photosynthetic bioreactor approach. Energy Conversion and Management 46(3): 403–420.

[12] Pires, J. C. M., Martins, F. G., Alvim-Ferraz, M. C. M. and Simões, M. 2011. Recent developments on carbon capture and storage: An overview. Chemical Engineering Research and Design 89(9): 1446–1460.

[13] Olajire, A. A. 2010. CO_2 capture and separation technologies for end-of-pipe applications—A review. Energy 35(6): 2610–2628.

[14] Chandel, M. K., Pratson, L. F. and Williams, E. 2010. Potential economies of scale in CO_2 transport through use of a trunk pipeline. Energy Conversion and Management 51(12): 2825–2834.

[15] Skovholt, O. 1993. CO_2 transportation system. Energy Conversion and Management 34(9): 1095–1103.

[16] Socolow, R. H. 2005. Can we bury global warming? Scientific American 293(1): 49–55.

[17] Rennie, A., Maxine C. Ackhurst, Gomersall, S., Pershad, H., Todd, A., Simon Forshaw, Stuart, Murray et al. 2009. Opportunities for CO_2 Storage around Scotland; An Integrated Strategic Research Study. SCCS.

[18] Forward, Y. 2008. A carbon capture and storage network for Yorkshire and Humber. An introduction to understanding the transportation of CO_2 from Yorkshire and Humber emitters into offshore storage sites.
[19] Morbee, J., Serpa, J. and Tzimas, E. 2012. Optimised deployment of a European CO_2 transport network. International Journal of Greenhouse Gas Control 7: 48–61.
[20] Rickard Svensson, Mikael Odenberger, Filip Johnsson and Lars Str€omberg. 2004. Transportation systems for CO_2—application to carbon capture and storage. Energy Conversion and Management 45(15): 2343–2353.
[21] McCoy, S. T. and Rubin, E. S. 2008. An engineering-economic model of pipeline transport of CO_2 with application to carbon capture and storage. International Journal of Greenhouse Gas Control 2(2): 219–229.
[22] Knoope, M. M. J., Ramírez, A. and Faaij, A. P. C. 2013. A state-of-the-art review of techno-economic models predicting the costs of CO_2 pipeline transport. International Journal of Greenhouse Gas Control 16: 241–270.
[23] Machteld van den Broek, Evelien Brederode, Andrea Ramírez, Leslie Kramers, Muriel van der Kuip, Ton Wildenborg et al. 2010. Designing a cost-effective CO_2 storage infrastructure using a GIS based linear optimization energy model. Environmental Modelling & Software 25(12): 1754–1768.
[24] Heddle, G., Herzog, H. and Klett, M. 2003. The Economics of CO_2 Storage. Massachusetts Institute of Technology, Laboratory for Energy and the Environment.
[25] Pipeline Infrastructure, C. 2010. An analysis of global challenges and opportunities. Final Report For International Energy Agency Greenhouse Gas Programm 27(04).
[26] Lanyu, G. A. O., Mengxiang, F. A. N. G., Hailong, L. I. and Jens HETLAND. 2011. Cost analysis of CO_2 transportation: case study in China. Energy Procedia 4: 5974–5981.
[27] Piessens, K., Laenen, B., Nijs, W., Mathieu, P., Baele, J. m., Hendriks, C. et al. 2008. Policy support system for carbon capture and storage. Relatório final para a Science for a Sustainable Development (SSD).

CHAPTER 4

Gas-Hydrate-Forming Reactors for Carbon Capture and Storage

Adeel Ur Rehman and *Bhajan Lal**

4.1 Introduction

Human-caused carbon emissions have increased dramatically since the preindustrial era due to rapid economic and population growth. Carbon emissions from burning fossil fuels and making cement have increased by 97% over the past four decades, from 5.0 GtC/yr (1977) to 9.9 GtC/yr (2017) [1]. In addition, land use changes like deforestation, turning forests into farmland, and urbanisation have resulted in a transfer of carbon from the land to the atmosphere at a rate of about 1.5 GtC per year. When anthropogenic carbon emissions are released into the atmosphere, they redistribute into three main segments: approximately 45 percent remain in the atmosphere, around 30 percent on land, and roughly 25 percent in the ocean. The land "sink" is the process by which plants and soils on land take in carbon dioxide from the atmosphere. Oceanic uptake occurs through CO_2 dissolving into the ocean. However, the ocean acidifies and the ocean chemistry shifts as a result of this process, which is bad for marine ecosystems. There is also the risk of re-emission of carbon from land and ocean sinks back into the atmosphere because of global warming, which could reduce the solubility of CO_2 in seawater [1], see Figure 1.

Since the initial identification of gas hydrates, researchers have designed a diverse range of tools for their examination. Implementing carbon capture and storage (CCS) based on gas hydrates necessitates

Chemical Engineering Department, Center for Carbon Capture, Storage and Utilization (CCUS), Institute of Sustainable Energy, Universiti Teknologi PETRONAS, Bandar Seri Iskandar, Perak, Malaysia.

* Corresponding author: bhajan.lal@utp.edu.my

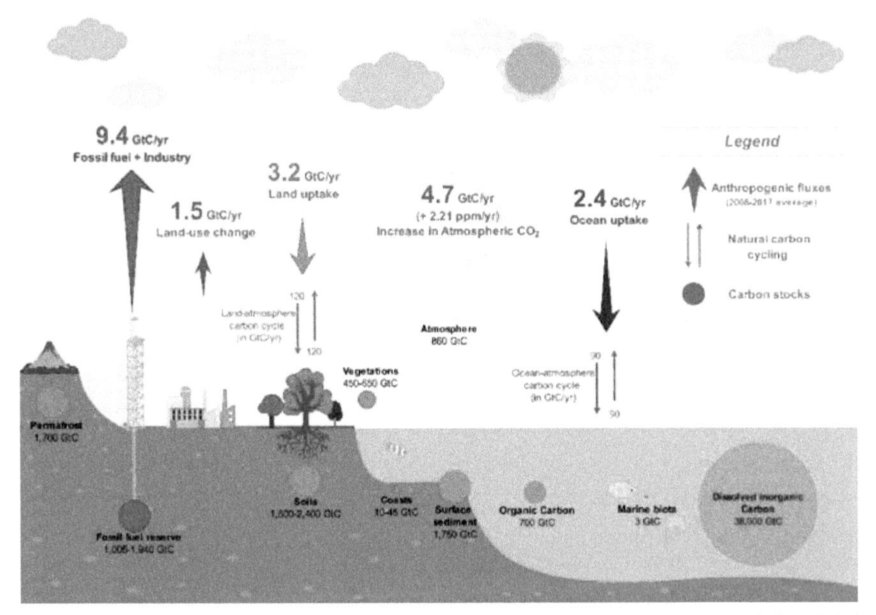

Figure 1. Anthropogenic activities' average ten-year (2008–2017) disruption of the global carbon cycle, depicted schematically with the active carbon cycles and carbon stocks [1].

a consistently operational unit with substantial capacity, capable of managing diverse compositions and flow rates.

Typical gas hydrate crystallizers and reactors were reviewed by Mori et al. [2]. Typically, gas hydrate crystallizers for studies on both thermodynamics and kinetics utilize stirrer-tanks, bubble columns, or spray towers. Additionally, rocking cells are employed for thermodynamic investigations [3]–[5] and the flow loops are used to simulate pipeline conditions [6]–[9]. Furthermore, there exist several distinctive crystallizers, most of which are presently safeguarded by patents. Within this section, various reactor configurations have been examined, highlighting their advantages and disadvantages for potential industrial use. However, this evaluation does not cover rocking cells and flow loops as these reactor types are not suitable for large-scale industrial gas hydrate production.

4.2 General Consideration on Gas Hydrate Reactor Design

Gas hydrate formation occurs at a slow pace and the duration for stable hydrate formation varies based on the prevailing conditions. Typically, experimental setups and laboratory reactors utilize batch-type formation for hydrates, whereas a continuous system is required for CO_2 capture, capable of managing substantial CO_2 gas volumes from or to the source. Numerous methods exist to hasten the rate of hydrate formation, making this technology viable for CO_2 capture. One approach to minimize the batch formation time involves employing hydrate formers, facilitating rapid hydrate formation and reducing batch duration. Implementing hydrate formers can involve multiple reactors, where one batch is forming while another reactor focuses on hydrate removal.

Semi-batch and continuous systems for hydrate formation can effectively capture bulk CO_2 quantities when integrated with the source plant. However, the challenge in semi-batch and continuous operations lies in extracting or removing hydrates from the reactor for dissociation. To address this issue, various designs of hydrate reactors have been utilized. Subsequent sections delve into discussions regarding these designs.

4.3 Modes of Reactor Operation

This section deals exclusively with the hydrate-forming gas, which may be either a single element or a mixture of two or more elements in a gaseous form, and with the liquid water that is fed into each reactor. Throughout the hydrate formation process, both the gas and liquid phases are kept within the reactor. This operation can occur in any of the following three modes:

4.3.1 Batch Hydrate Formation Process

a) In this closed-system batch operation, the hydrate-forming gas and liquid water are introduced into the reactor at the outset, and the reactor remains sealed throughout the process. No additional gas or water is introduced or removed during the operation. As hydrate formation proceeds, the gas molecules become incorporated into the hydrate structure, causing a continuous decline in pressure within the reactor. This pressure decline is attributed to the diminishing number of gas molecules in the gas phase. As hydrate forms, gas molecules become trapped within the hydrate lattice, effectively removing them from the gas phase. This leads to a reduced overall gas density and, consequently, a lower pressure [10].

b) The rate of pressure decrease is influenced by various factors, including the composition of the hydrate-forming gas, the prevailing temperature and pressure conditions, and the available surface area for hydrate formation. Generally, the pressure decrease is more pronounced for gases with higher hydrate-forming capacities and at lower temperatures and pressures [11].

c) By monitoring the continuous pressure drop during this batch operation, the progress of hydrate formation can be tracked. As the pressure approaches a stable level, it indicates that hydrate formation is nearing completion.

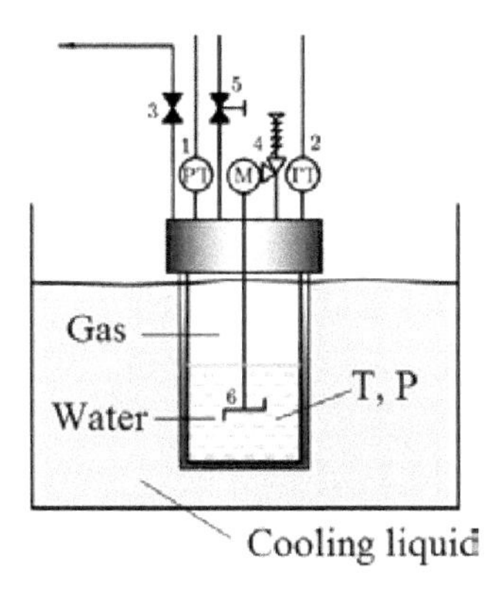

1. Inlet pressure transmitter
2. Inlet temperature transmitter
3. Gas outlet
4. Motor
5. Gas inlet
6. Stirrer

Figure 2. Batch-wise hydrate forming reactor setup [12].

d) This type of batch operation (shown in Figure 2) is frequently employed in laboratory studies of hydrate formation kinetics and for small-scale hydrate production. It offers a relatively straightforward and controlled method for investigating hydrate formation phenomena [12].

4.3.2 Semi-batch or Semi-continuous Hydrate Formation

a) In this semi-batch operation, the hydrate-forming gas is continuously fed into the reactor at a rate that matches the gas consumption during hydrate formation. This controlled gas injection maintains a nearly constant pressure within the reactor, ensuring stable conditions for hydrate growth.

b) Unlike a batch operation where the gas is introduced once and the reactor remains sealed, this semi-batch approach allows for continuous gas replenishment, compensating for the gas molecules that become enmeshed in the hydrate structure. This strategy effectively maintains the pressure equilibrium within the reactor, preventing pressure fluctuations that could hinder hydrate formation.

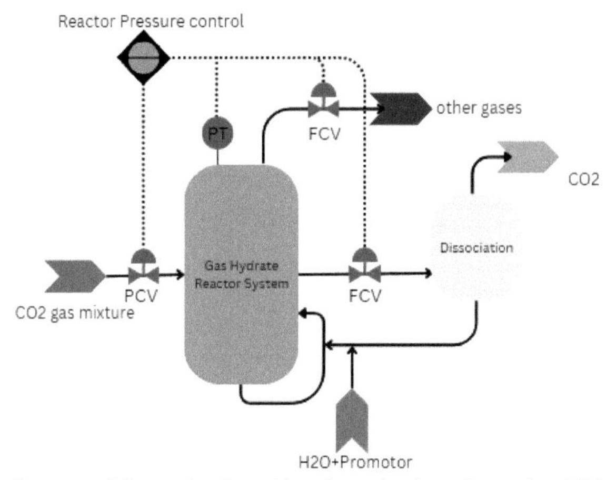

Figure 3. Schematic of semi-batch gas hydrate formation [13].

c) The rate of gas introduction is carefully regulated to match the rate of hydrate formation. By closely monitoring the pressure and gas consumption, the semi-batch operation ensures a steady supply of gas molecules for hydrate growth while maintaining a stable pressure environment.

d) This semi-batch operation is particularly useful for larger-scale hydrate production processes, where maintaining constant pressure is crucial for efficient hydrate formation [13]. It offers a controlled and efficient method for producing hydrates in industrial settings, see Figure 3.

4.3.3 *Continuous Hydrate Formation*

a) In this continuous process, both the hydrate-forming gas and liquid water are continuously introduced into the reactor, while a slurry of formed hydrate and liquid phase is continuously removed. This continuous flow of material maintains a near-constant pressure and ratio of gas, liquid, and hydrate within the reactor, ensuring optimal conditions for efficient hydrate production.

b) Similar to a river flowing steadily, the continuous input of gas and water and the removal of hydrate slurry maintain a dynamic equilibrium within the reactor. This controlled flow of reactants and products ensures that the reaction proceeds smoothly and efficiently, maximizing hydrate production.

c) The continuous process is particularly well-suited for large-scale industrial applications where consistent and high-volume production of hydrates is desired. It offers a reliable and efficient method for

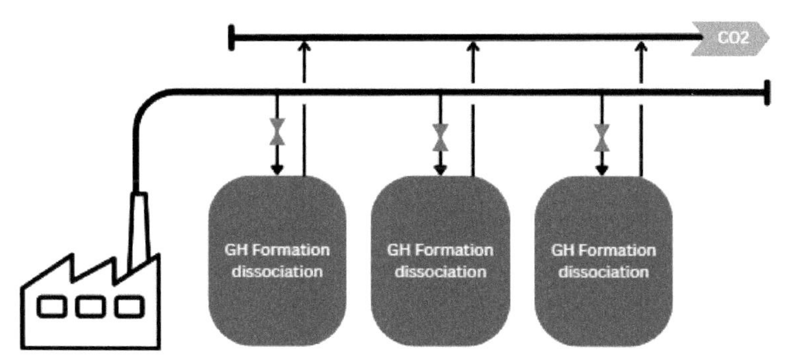

Figure 4. Schematic of continuous CO_2 sequestration by gas hydrate system [7].

producing hydrates in large quantities, meeting the demands of various industries. Here's a Figure 4 to illustrate the continuous hydrate formation process.

4.4 Systems for Gas-Liquid Interfacing and Thermal Management

Two essential conditions for reactors to achieve high hydrate formation rates are: (1) ensuring effective mixing between the gas that forms hydrates and the liquid water, and (2) employing efficient cooling systems to remove the heat generated during hydrate formation [7]. Inadequate gas-liquid mixing and inefficient heat dissipation render reactor designs unsuitable for industrial hydrate production.

The most common way to satisfy the first requirement is to disperse either the gas or liquid water into the other, which calls for mechanical effort in each reactor by means of agitator or stirrer. Injecting individual gas bubbles into an aqueous phase is a typical technique for dispersing the gas phase [3–5, 8], or dispersing liquid water by spraying or sprinkling into a continuous gas phase [9–13], substantially amplifies the gas-liquid interfacial area. This increase facilitates gas dissolution into the aqueous phase, thereby encouraging hydrate formation at or near the interface. The external work required to disperse either fluid phase into the other can be disregarded when a surfactant is introduced to the lower portion of the tank-shaped reactor, leading to a spontaneous growth of a capillarity-driven hydrate layer along the reactor wall above the still aqueous pool's free surface (see example [14–16]). Reducing the wall-surface-to-volume ratio as the reactor size increases can be counteracted by vertically hanging metal plates within the reactor from its upper part. These plates are immersed in the aqueous pool, functioning as supplementary locations for hydrate-layer development while simultaneously operating as cooling fins. This dual role helps dissipate heat from the reactor to its external surroundings [15, 17, 18]. Another potential approach to seemingly bypass the need for mechanical work in gas-liquid mixing involves employing fixed-bed reactors. This reactor type utilizes packing materials like silica gels, silica sand, polyurethane foam, among others. It has been effectively utilized in recent laboratory experiments examining the viability of CO_2 capture from flue gas using hydrate-based methods [19–21]. In this scenario, the filling material within each reactor can be likened to a static mixer, facilitating thorough gas-liquid mixing while incurring an added pressure drop in the gas flow. This process indirectly involves the utilization of mechanical work.

Regarding the second requirement, multiple techniques and equipment designs for heat dissipation have been created and put to practical tests. Common cooling approaches for hydrate formation reactors can be broadly classified into two main categories: direct reactor cooling and indirect cooling techniques [4, 5, 10] or a target cooling, decreasing the temperature of the reactor at a specific place [22], alternatively, liquid water (or a hydrate slurry) can be circulated through an external loop equipped with a heat exchanger for cooling [8, 9, 11]. The external loop cooling method can be extended to include a separate loop for circulating a hydrophobic liquid coolant, in addition to water circulation. This coolant is cooled to a temperature significantly lower than the freezing point of water, resulting in substantial cooling energy transfer to the reactor [23–25]. Both approaches can be employed simultaneously to cool a singular reactor unit [3].

4.5 Advanced Reactor Designs and Scaling for Gas Hydrate Production

A multifunctional hydrate production reactor was developed by Zhichao Liu and colleagues to simulate gas-water-sand production and evaluate the multifield and multiphase processes associated with hydrate production. The apparatus, described, is designed to operate under high pressures of up to 25 MPa and a wide temperature range from –20°C to 50°C, making it capable of replicating the complex conditions of hydrate reservoirs. The setup includes six primary subsystems: a fluid supplement unit, a multifunction chamber, an axial loading unit, a hydrate production unit, a temperature control unit, and a data recorder. The fluid supplement unit, comprising a gas source, liquid source, and gas-liquid mixing tube, controls the pore pressure during experiments. The multifunction chamber, made of 316 L stainless steel and tempered glass, simulates the hydrate reservoir zone around a wellbore and is equipped with sensors for measuring temperature, pressure, deformation, stress, wave propagation, and electrical resistivity. The axial loading unit applies axial stress to the specimen to simulate overburden stress using a servo pump, while the hydrate production unit, including a gas-liquid-solid phase separator and back-pressure servo pump, manages the formation and dissociation of hydrates and separates the produced gas, water, and sand. The apparatus operates by packing water and sand in the chamber, injecting gas to form hydrates, and triggering dissociation by releasing back pressure. Hydrate formation is indicated by pressure drop and temperature peaks, and the produced gas is measured [48].

The formation and dissociation of gas hydrates have garnered significant interest due to their applications in various fields, including natural gas storage, carbon dioxide sequestration, and secondary refrigeration systems. The efficient production of gas hydrates necessitates an understanding of the complex processes involved, including nucleation, growth, and dissociation kinetics, as well as the impact of different reactor designs and operational conditions. Research into gas hydrate reactors and crystallizers has led to the development of various innovative systems aimed at optimizing hydrate formation rates and gas storage capacities. These systems are designed to address the inherent challenges associated with hydrate production, such as mass transfer limitations, heat management, and the need for high-pressure and low-temperature conditions. The diversity of reactor designs reflects the multifaceted nature of hydrate research, encompassing techniques such as stirring, spraying, bubbling, and the application of electrostatic fields to enhance gas-water interactions. The design and operational principles of these reactors, highlighting their specific applications and the efficiencies achieved in hydrate formation. By examining various reactor types, including mixed arrangement reactors, continuously stirred tank reactors (CSTRs), and microfluidic systems, we aim to elucidate the mechanisms that drive efficient gas hydrate production [48].

In the study by Liu et al.'s apparatus allows for detailed monitoring and modeling of hydrate production processes, providing valuable insights into the mechanical, hydraulic, and electrical behaviors of hydrate reservoirs. An integrated case study on a synthetic specimen with 12.8% hydrate saturation demonstrated the apparatus's capabilities, highlighting key findings such as the significant production of gas and water during the early stages of depressurization and the impact of sand production on the specimen's structural stability. These results support the development of optimized field strategies for safe and economical gas production from hydrates, contributing to a better understanding of the coupled behaviors during hydrate production [48].

Xin Yang and colleagues, a three-dimensional middle-size reactor was developed to investigate gas production from methane hydrate-bearing sand via hot-water cyclic injection. This reactor was designed to simulate the entire gas production process, which includes three main steps: injecting hot water, closing the well, and producing gas. The experimental apparatus comprises four main sections: the reacting system, water injection system, gas production system, and monitor and control generated system (MCGS). The reacting system is centered around a high-pressure reactor with an inner diameter of 300 mm and an effective height of 100 mm, capable of withstanding pressures up to 16 MPa. The reactor is immersed in a water-ethylene glycol solution to maintain a constant temperature with high precision. Sixteen thermocouples are inserted into the reactor to monitor temperature variations throughout the experiment. Hot water, prepared by a water heater with thermostatic control, is injected into the reactor through a well using a metering pump. The gas production system includes a gas-water separator, filter, back-pressure regulator, and mass flow transducers to measure the produced gas and water. The MCGS records temperature, pressure, and gas flow rate data. The findings highlight the importance of temperature distribution and its impact on gas production efficiency, offering a deeper understanding of the thermal stimulation method in hydrate reservoirs. This reactor setup provides a robust platform for optimizing gas production strategies from hydrate-bearing sediments, contributing significantly to the field of energy production from gas hydrates [49].

Martinez de Baños et al.'s study highlights the potential of droplet-based millifluidic methods for investigating the kinetics of hydrate crystallization. The apparatus consists of a spiral capillary tube containing regularly spaced water droplets dispersed in a hydrate-former phase, specifically cyclopentane (CP), which forms hydrates at ambient pressure and temperatures below 7.2°C. The capillary tube is made of perfluoroalkoxy-alkane (PFA) with an outer diameter of 1.59 mm and an inner diameter of 1 mm, connected to syringe pumps for co-injecting water and CP to create a train of immobile water droplets. The apparatus is equipped with a temperature control system using a Peltier element and a cooling circuit with water and glycol to maintain precise thermal conditions. The capillary tube is sandwiched between a sapphire window and an aluminum holder, allowing for temperature regulation and visualization through an imaging system consisting of a camera mounted on a zoom system. The apparatus allows for the detailed observation of hydrate nucleation and growth within the water droplets [50].

Chen and Hartman's study demonstrates the effectiveness of using a microfluidic approach to investigate methane hydrate dissociation. The system provides exceptional control over experimental conditions, allowing for detailed analysis of the intrinsic and heat-transfer kinetics of hydrate dissociation. This innovative setup is specifically designed to provide detailed insights into the dissociation process by controlling temperature and pressure conditions with high precision. The microfluidic system consists of microchannels etched into a silicon wafer with dimensions of 400 × 400 μm, bonded to a Pyrex glass wafer to create three-dimensional microchannels. The entire system is integrated with a thermoelectric cooling device that allows for precise temperature control, with the capability to adjust the temperature in increments as small as 0.1 K. The system operates under pressures up to 80 bar, simulating the conditions necessary for methane hydrate formation and dissociation [51].

Xiao-Sen Li et al., studied the development and application of a Pilot-Scale Hydrate Simulator (PHS) are described to investigate gas production from methane hydrate in porous media using the huff and puff method. This reactor is a three-dimensional pressure vessel with a capacity of 117.8 liters, designed to withstand pressures up to 30 MPa. The PHS is used to simulate the conditions of methane hydrate reservoirs and to study the efficiency of the huff and puff method, also known as cyclic steam stimulation, for gas production. The PHS is a cylindrical pressure vessel with dimensions of 0.60 meters in length and 0.50 meters in diameter. It is equipped with a complex well system consisting of 27 vertical wells and 3 horizontal wells, distributed in three horizontal layers (A–A, B–B, and C–C) that equally divide the vessel into four parts. The vertical wells are used for the injection and production of hot water, gas, and water, while the horizontal wells facilitate the distribution of fluids within the vessel. The apparatus also includes 49 thermometers and resistance ports evenly distributed across each layer, allowing for precise temperature and resistance measurements. This reactor setup offers a robust platform for optimizing gas production strategies from hydrate-bearing sediments and contributes significantly to the understanding of hydrate dissociation mechanisms. This reactor, with its ability to replicate the complex conditions of natural hydrate reservoirs, supports the development of efficient and economical gas production methods, making it a valuable tool in the field of energy production from gas hydrates [52].

Li et al.'s study using the micro-packed bed reactor provides valuable insights into methane hydrate formation and dissociation in porous media. This reactor aims to simulate the porous conditions of sediments and provide direct visualization of phase behavior, morphology, interfacial phenomena, and fluid migration in pore spaces. The reactor is specifically designed for methane hydrate research, addressing both formation and depressurization-induced dissociation behaviors. The micro-packed bed reactor is constructed from polymethyl methacrylate (PMMA) and has a depth of 1 mm. It is filled with 1 mm diameter glass beads to simulate the porous environment of hydrate sediments. The dissociation of methane hydrate is induced by a continuous depressurization process, where fluids are extracted at a constant rate. The reactor's pressure is reduced in three stages: an initial linear decrease with no hydrate dissociation, a second stage where hydrate dissociation occurs with gas release, and a final stage where complete dissociation is achieved, accompanied by significant gas-water migration. The study highlights the impact of fluid extraction rate on the dissociation process, with higher rates leading to more severe hydrate shell cracking and reformation phenomena. The reactor is integrated with a high-pressure jacket, temperature control system, and microscopic imaging system. Methane and water are injected into the reactor using high-pressure syringe pumps, and the system operates under controlled temperature and pressure conditions to form and dissociate methane hydrates [53].

The study by Bai et al. effectively demonstrates the synergistic effects of surface tension reduction and impinging flow on methane hydrate formation. The custom-designed imping flow reactor significantly enhances the kinetics of hydrate formation, providing valuable insights for the development of efficient hydrate accelerators and industrial applications of hydrate technology. The experimental setup includes a liquid continuous impinging stream (LIS) reactor made of stainless steel with a capacity of 2 liters. The reactor features two reverse impellers with blade angles of 45° installed on an agitator shaft to create two opposite fluid collisions, thereby increasing fluid turbulence. The reactor is placed in a temperature-controlled thermostat, capable of maintaining temperatures between 248.15 K and 358.15 K with an accuracy of ±0.05 K. Temperature is measured using a Pt 100 sensor, and pressure is monitored with a CYB-20S sensor. The reactor also includes a surface tension tester for measuring the gas-liquid interfacial tension, employing a BZY-101 tensiometer. This novel reactor design aims to enhance the understanding and efficiency of methane hydrate synthesis, focusing on the role of surface tension and the impact of impinging flow on hydrate formation kinetics [54].

In the study by Véronique Osswald and colleagues, a novel experimental setup was developed to monitor CO_2 hydrate slurry crystallization by determining the heat flux rate in a jacketed reactor. This reactor aims to provide accurate insights into the kinetics of CO_2 hydrate formation, which is crucial for their application in secondary refrigeration systems due to their high energy density. The reactor is a customized stainless steel jacketed vessel with a magnetic drive coupling motor, capable of withstanding pressures up to 3.8 MPa and an internal volume of 1.4 liters. It features a cooling jacket, which is controlled by an external heating/cooling unit (Julabo FP50-HE) that maintains the temperature with a stability of 0.01°C. The cooling jacket is equipped with a thermopile consisting of six T-type thermocouples to measure the temperature difference between the inlet and outlet of the cooling jacket, allowing for direct heat flux measurements. The reactor can be fitted with either a 3-blade propeller or a Dispersimax turbine, with stirring speeds adjustable from 100 to 800 rpm. The reactor setup provides a robust platform for optimizing the formation process, crucial for industrial applications of CO_2 hydrates in refrigeration systems [55].

In the study by Phillip Szymcek and colleagues, a pilot-scale continuous-jet hydrate reactor (CJHR) was developed and tested for the production of gas hydrates, specifically targeting CO_2 hydrate formation for ocean carbon sequestration. The CJHR is a three-phase pilot-scale reactor comprising a 72-liter high-pressure vessel capable of operating at pressures equivalent to intermediate ocean depths (1100–1700 meters). The reactor is constructed from stainless steel and features a headpiece with interchangeable inputs for hydrate-forming species and water, connected to a 45.7-cm stainless steel tube. The system includes various capillary injector designs to enhance the dispersion of reactants and optimize hydrate formation. These injectors range from single-capillary to multiple-capillary configurations, designed to break the flow into small droplets and maximize interfacial area for efficient mass transfer. The hydrate formation process in the CJHR involves injecting CO_2 and water through separate inlets into the reactor, where one reactant is dispersed into the other using the capillary injectors. The dispersion creates small droplets, facilitating better contact between CO_2 and water, thereby overcoming mass transfer barriers. The system operates under controlled temperatures and pressures to simulate the conditions required for CO_2 hydrate formation [56].

Rossi et al.'s study demonstrates the effectiveness of a novel reactor design for the rapid and continuous production of methane hydrates. The reactor is a high-pressure cylindrical vessel made of AISI 304 stainless steel, with an internal diameter of 200 mm, an internal length of 800 mm, and a total internal volume of 25 liters. It is equipped to handle pressures up to 120 bar and includes a safety valve for protection. The reactor features six spray nozzles at the top, which inject water into the methane-filled reactor, creating a fine mist that significantly increases the interfacial area between the water and methane, thus enhancing the hydrate formation rate.

A system for heat removal is integrated into the reactor to manage the exothermic heat released during hydrate formation and maintain a constant temperature. Additionally, a screw conveyor inside the reactor helps in the recovery and unloading of solid hydrates. The reactor also includes various ports for temperature sensors and methane and water injection lines, with temperature control provided by an external thermostatic bath filled with an ethylene glycol-water solution. The reactor operates by first filling it with methane gas to the desired pressure and then cooling it using liquid nitrogen. Water, with or without surfactants like sodium dodecyl sulfate (SDS), is pressurized and injected into the reactor through the spray nozzles, creating a fine mist that interacts with the methane gas to form hydrates. The temperature and pressure within the reactor are carefully monitored and controlled to ensure optimal hydrate formation conditions. This reactor is designed to maximize the efficiency of gas hydrate formation by enhancing the interfacial area between reactants and minimizing mass transfer barriers and thermal effects. The reactor's spray nozzle system and integrated heat removal capabilities significantly enhance the formation rate and efficiency of hydrates, making it a promising solution for industrial applications in gas storage and transportation. The use of surfactants like SDS further improves the performance, reducing the time required for hydrate formation and increasing methane capture [57].

4.6 Techno-economic Limitations to Hydrate-Based CCS

Numerous research efforts have focused on the technical advancements and improvements in hydrate-based CO_2 capture methods, yet the economic implications of these technologies have received comparatively less attention [34]. A study by Babu et al. [35] conducted a comprehensive comparison of hydrate-based CO_2 capture with established methods such as SELEXOL and SIMTECHE processes. Their evaluation was based on a coal gasifier generating 500 MW of electricity. The study revealed that a twin-stage Tetrahydrofuran (THF) promoter system for hydrate-based CO_2 capture resulted in an estimated electricity cost of 6.13 and 6.24 cents/kWh.

However, due to the relative novelty of hydrate-based CO_2 sequestration technologies, particularly in geological formations, there is a dearth of economic data available to comprehensively assess their feasibility. This section delves into the cost estimation of various components involved in hydrate-based carbon capture and storage (CCS) systems. A study by Duc et al. [36] explored the cost of a hydrate-based CO_2 capture technology utilizing TBAB as a promoter to capture CO_2 from flue gases generated by the steel industry. Their findings indicated that the costs associated with reducing CO_2 emissions ranged from $20 to $40 per metric ton of CO_2, with variations depending on operational parameters and feed gas quality. Additionally, customer acquisition costs for hydrate-based CO_2 capture were reported to be lower compared to the standard absorption-based CO_2 capture approach, which typically incurs costs between $40 and $100 per ton of CO_2 captured [37].

Spencer et al. [38] investigated the applicability of hydrate-based CO_2 capture methods for syngas applications. Their research demonstrated that the energy cost of using hydrates to capture carbon dioxide is substantially lower, ranging from 50 to 75%, compared to absorption methods. This finding highlights the potential of hydrate-based capture as a sustainable and cost-effective alternative to absorption-based capture.

Given the relatively nascent stage of development of hydrate-based CO_2 sequestration technologies, extensive research is still required to refine existing methods and ensure their continued adoption and implementation within government and industry sectors. This ongoing research will be crucial for establishing hydrate-based CO_2 capture as a viable and economically advantageous solution for mitigating carbon emissions and addressing climate change concerns.

References

[1] Zheng, J., Chong, Z. R., Qureshi, M. F. and Linga, P. 2020. Carbon dioxide sequestration via gas hydrates: a potential pathway toward decarbonization. Energy and Fuels 34(9): 10529–10546, doi: 10.1021/acs.energyfuels.0c02309.

[2] Mori, Y. H. 2015. On the scale-up of gas-hydrate-forming reactors: The case of gas-dispersion-type reactors. Energies 8(2): 1317–1335, doi: 10.3390/en8021317.

[3] Sundramoorthy, J. D., Sabil, K. M., Lal, B. and Hammonds, P. 2015. Catastrophic crystal growth of clathrate hydrate with a simulated natural gas system during a pipeline shut-in condition. Cryst. Growth Des. 15(3): 1233–1241.

[4] Saberi, A., Alamdari, A., Rasoolzadeh, A. and Mohammadi, A. H. 2020. Insights into kinetic inhibition effects of MEG, PVP, and L-tyrosine aqueous solutions on natural gas hydrate formation. Pet. Sci., no. 0123456789, doi: 10.1007/s12182-020-00515-0.

[5] Daraboina, N., Malmos, C. and von Solms, N. 2013. Synergistic kinetic inhibition of natural gas hydrate formation. Fuel 108: 749–757.

[6] Jerbi, S., Delahaye, A., Fournaison, L. and Haberschill, P. 2010. Characterization of CO_2 hydrate formation and dissociation kinetics in a flow loop. Int. J. Refrig. 33(8): 1625–1631.

[7] Sarshar, M., Fathikalajahi, J. and Esmaeilzadeh, F. 2010. Experimental and theoretical study of gas hydrate formation in a high-pressure flow loop. Can. J. Chem. Eng. 88(5): 751–757.

[8] Fu, W., Wang, Z., Yue, X., Zhang, J. and Sun, B. 2019. Experimental study of methane hydrate formation in water-continuous flow loop. Energy & Fuels 33(3): 2176–2185.

[9] Lv, X. et al. 2019. Experimental study of growth kinetics of CO_2 hydrates and multiphase flow properties of slurries in high pressure flow systems. RSC Adv. 9(56): 32873–32888.

[10] Hassan, M. H. A., Sher, F., Zarren, G., Suleiman, N., Tahir, A. A. and Snape, C. E. 2020. Kinetic and thermodynamic evaluation of effective combined promoters for CO2 hydrate formation. J. Nat. Gas Sci. Eng. 78: 103313.

[11] Thoutam, P., Rezaei Gomari, S., Chapoy, A., Ahmad, F. and Islam, M. 2019. Study on CO_2 hydrate formation kinetics in saline water in the presence of low concentrations of CH_4. ACS Omega 4(19): 18210–18218.

[12] Bozorgian, A., Arab Aboosadi, Z., Mohammadi, A., Honarvar, B. and Azimi, A. 2020. Evaluation of the effect of nonionic surfactants and TBAC on surface tension of CO_2 gas hydrate. J. Chem. Pet. Eng. 54(1): 73–81.

[13] Wang, X., Zhang, F. and Lipiński, W. 2020. Research progress and challenges in hydrate-based carbon dioxide capture applications. Appl. Energy 269: 114928.

[14] Khan, M. S., Lal, B., Sabil, K. M. and Ahmed, I. 2019. Desalination of seawater through gas hydrate process: an overview. J. Adv. Res. Fluid Mech. Therm. Sci. 55(1): 65–73.

[15] Babu, P., Kumar, R. and Linga, P. 2014. Unusual behavior of propane as a co-guest during hydrate formation in silica sand: Potential application to seawater desalination and carbon dioxide capture. Chem. Eng. Sci. 117: 342–351.
[16] Partoon, B., Sabil, K. M., Roslan, H., Lal, B. and Keong, L. K. 2016. Impact of acetone on phase boundary of methane and carbon dioxide mixed hydrates. Fluid Phase Equilib. 412: 51–56, doi: 10.1016/j.fluid.2015.12.027.
[17] Khan, M. S., Lal, B., Partoon, B., Keong, L. K., Bustam, A. B. and Mellon, N. B. 2016. Experimental evaluation of a novel thermodynamic inhibitor for CH_4 and CO_2 Hydrates. Procedia Eng. 148: 932–940, doi: 10.1016/j.proeng.2016.06.433.
[18] Cha, J. H. and Seol, Y. 2013. Increasing gas hydrate formation temperature for desalination of high salinity produced water with secondary guests," ACS Sustain. Chem. Eng. 1(10): 1218–1224, doi: 10.1021/sc400160u.
[19] Mori, H. 2003. Recent advances in hydrate-based technologies for natural gas storage—a review. 化工学报(z1): 1–17.
[20] Luo, Y.-T., Zhu, J.-H., Fan, S.-S. and Chen, G.-J. 2007. Study on the kinetics of hydrate formation in a bubble column. Chem. Eng. Sci. 62(4): 1000–1009.
[21] Hashemi, S., Macchi, A. and Servio, P. 2009. Gas–liquid mass transfer in a slurry bubble column operated at gas hydrate forming conditions. Chem. Eng. Sci. 64(16): 3709–3716.
[22] Xu, C.-G., Li, X.-S., Lv, Q.-N., Chen, Z.-Y. and Cai, J. 2012. Hydrate-based CO_2 (carbon dioxide) capture from IGCC (integrated gasification combined cycle) synthesis gas using bubble method with a set of visual equipment. Energy 44(1): 358–366.
[23] Fukumoto, K., Tobe, J., Ohmura, R. and Mori, Y. H. 2001. Hydrate formation using water spraying in a hydrophobic gas: a preliminary study. Am. Inst. Chem. Eng. AIChE J. 47(8): 1899.
[24] Ohmura, R., Kashiwazaki, S., Shiota, S., Tsuji, H. and Mori, Y. H. 2002. Structure-I and structure-H hydrate formation using water spraying. Energy & Fuels 16(5): 1141–1147.
[25] Li, G., Liu, D., Xie, Y. and Xiao, Y. 2010. Study on effect factors for CO_2 hydrate rapid formation in a water-spraying apparatus. Energy & Fuels 24(8): 4590–4597.
[26] Kikuo, N. 2006. Hydrate Production Equipment. JP 2006-111774 A.
[27] Lucia, B. et al. 2014. Experimental investigations on scaled-up methane hydrate production with surfactant promotion: Energy considerations. J. Pet. Sci. Eng. 120: 187–193.
[28] Partoon, B., Sabil, K. M., Lau, K. K., Lal, B. and Nasrifar, K. 2018. Production of gas hydrate in a semi-batch spray reactor process as a means for separation of carbon dioxide from methane. Chem. Eng. Res. Des. 138: 168–175, doi: 10.1016/j.cherd.2018.08.024.
[29] Heinemann, R. F., Huang, D. D.-T., Long, J. and Saeger, R. B. 2000. Process for making gas hydrates. US6028234A, Feb. 22, 2000.
[30] Szymcek, P., McCallum, S. D., Taboada-Serrano, P. and Tsouris, C. 2008. A pilot-scale continuous-jet hydrate reactor. Chem. Eng. J. 135(1–2): 71–77.
[31] Iwasaki, T., Katoh, Y., Nagamori, S., Takahashi, S. and Oya, N. 2005. Continuous natural gas hydrate pellet production (NGHP) by process development unit (PDU).
[32] Lozada Garcia, N. et al. 2023. Characteristics of continuous CO_2 hydrate formation process using a NetMIX reactor. Chem. Eng. Sci. 280(May, 2023), doi: 10.1016/j.ces.2023.119023.
[33] Filarsky, F., Wieser, J. and Schultz, H. J. 2021. Rapid gas hydrate formation—evaluation of three reactor concepts and feasibility study. Molecules 26(12), doi: 10.3390/molecules26123615.
[34] Nguyen, N. N., La, V. T., Huynh, C. D. and Nguyen, A. V. 2022. Technical and economic perspectives of hydrate-based carbon dioxide capture. Appl. Energy 307: 118237.
[35] Babu, P., Linga, P., Kumar, R. and Englezos, P. 2015. A review of the hydrate based gas separation (HBGS) process for carbon dioxide pre-combustion capture. Energy 85: 261–279.

[36] Duc, N. H., Chauvy, F. and Herri, J.-M. 2007. CO_2 capture by hydrate crystallization–A potential solution for gas emission of steelmaking industry. Energy Convers. Manag. 48(4): 1313–1322.
[37] Dashti, H. and Lou, X. 2018. Gas hydrate-based CO_2 separation process: quantitative assessment of the effectiveness of various chemical additives involved in the process. In TMS Annual Meeting & Exhibition, pp. 3–16.
[38] Arora, V., Saran, R. K., Kumar, R. and Yadav, S. 2019. Separation and sequestration of CO_2 in geological formations. Mater. Sci. Energy Technol. 2(3): 647–656.
[39] Partoon, B. and Javanmardi, J. 2013. Effect of mixed thermodynamic and kinetic hydrate promoters on methane hydrate phase boundary and formation kinetics. J. Chem. Eng. Data 58(3): 501–509.
[40] Sabil, K. M., Witkamp, G.-J. and Peters, C. J. 2009. Phase equilibria of mixed carbon dioxide and tetrahydrofuran hydrates in sodium chloride aqueous solutions. Fluid Phase Equilib. 284(1): 38–43.
[41] Partoon, B., Malik, S. N. A., Azemi, M. H. and Sabil, K. M. 2013. Experimental investigations on the potential of SDS as low-dosage promoter for carbon dioxide hydrate formation. Asia-Pacific J. Chem. Eng. 8(6): 916–921.
[42] Seo, Y.-T. and Lee, H. 2001. Multiple-phase hydrate equilibria of the ternary carbon dioxide, methane, and water mixtures. J. Phys. Chem. B 105(41): 10084–10090.
[43] Clarke, M. A. and Bishnoi, P. R. 2005. Determination of the intrinsic kinetics of CO_2 gas hydrate formation using *in situ* particle size analysis. Chem. Eng. Sci. 60(3): 695–709.
[44] Murakami, T., Kuritsuka, H., Fujii, H. and Mori, Y. H. 2009. Forming a structure-H hydrate using water and methylcyclohexane jets impinging on each other in a methane atmosphere. Energy Fuels 23: 1619–1625.
[45] West, O. R., Tsouris, C., Lee, S., McCallum, S. D. and Liang, L. 2003. Negatively buoyant CO(2)-hydrate composite for ocean carbon sequestration. Am. Inst. Chem. Eng. AIChE J. 49: 283.
[46] Szymcek, P., McCallum, S. D., Taboada-Serrano, P. and Tsouris, C. 2008. A pilot-scale continuous-jet hydrate reactor. Chem. Eng. J. 135: 71–77.
[47] Tsouris, C., Szymcek, P., Taboada-Serrano, P., McCallum, S. D., Brewer, P., Peltzer, E. et al. 2007. Scaled-up ocean injection of CO_2–hydrate composite particles. Energy Fuels 21: 3300–3309.
[48] Zhichao Liu, Yingjie Zhao, Guocai Gong, Wei Hu, Zhun Zhang and Fulong Ning. Dec 2022. A novel apparatus for modeling the geological responses of reservoir and fluid-solid production behaviors during hydrate production.
[49] Yang, X., Sun, C.-Y., Yuan, Q., Ma, P.-C. and Chen, G.-J. 2010. Experimental study on gas production from methane hydrate-bearing sand by hot-water cyclic injection. Energy & Fuels 24(11).
[50] Maria Lourdes Martinez de Baños, Odile Carrier, Patrick Bouriat and Daniel Broseta. Feb 2015. Droplet-based millifluidics as a new tool to investigate hydrate crystallization: Insights into the memory effect.
[51] Weiqi Chen and Ryan L. Hartman. October 2018. Methane hydrate intrinsic dissociation kinetics measured in a microfluidic system by means of *in situ* raman spectroscopy. Energy Fuels 32(11): 11761–11771.
[52] Li, X.-S., Yang, B., Li, G., Li, B., Zhang, Y. and Chen, Z.-Y. 2012. Experimental study on gas production from methane hydrate in porous media by huff and puff method in Pilot-Scale Hydrate Simulator. Fuel 94: 486–494.
[53] Xingxun Li, Ming Liu, Qingping Li, Weixin Pang, Guangjin Chen and Changyu Sun. Jan 2023. Visual study on methane hydrate formation and depressurization-induced methane hydrate dissociation processes in a micro-packed bed reactor. Volume 23.

[54] Jing Bai, Yanqing Zhang, Yanhui Wang, Xianyun Wei, Chenxu Qiu, Chun Chang et al. 2024. Effects of surface tension on the kinetics of methane hydrate formation with APG additive in an impinging stream reactor. 1 May 2024, 363: 130889.
[55] Véronique Osswald, Pascal Clain, Laurence Fournaison and Anthony Delahaye. 2023. Experimental monitoring of CO_2 hydrate slurry crystallization by heat flux rate determination in a jacketed reactor. International Journal of Heat and Mass Transfer 15 December 2023, 217: 124665.
[56] Szymcek, P., McCallum, S. D., Taboada-Serrano, P. and Tsouris, C. 2008. A pilot-scale continuous-jet hydrate reactor. Chemical Engineering Journal 135(1-2): 71–77.
[57] Federico Rossi, Mirko Filipponi and Beatrice Castellani. 2012. Investigation on a novel reactor for gas hydrate production. Applied Energy November 2012, 99: 167–172.

CHAPTER 5

Advancements and Challenges in Techno-Economic Analysis of CO_2 Capture Technologies

*Anipeddi Manjusha** and *Bhajan Lal*

5.1 Introduction

Amidst the critical issue of climate change, the exploration of carbon dioxide (CO_2) utilization has become a focal point for industry, academic institutions, and policymakers. This surge of interest has catalyzed the development of a plethora of promising technologies for converting CO_2 into useful chemicals, fuels, and minerals. However, the term "promising technology" is subjective and does not imply a rigorous, systematic evaluation. It is crucial, therefore, to employed methodologies such as Techno-Economic Assessment (TEA) and Life Cycle Assessment (LCA) to steer research and development effectively toward the path of commercial viability [1]. TEA provides a structured approach for examining the technical and economic merits of processes, products, or services, while LCA quantifies the environmental impacts throughout their lifecycle.

Nonetheless, the lack of standardization in the application of TEA and LCA methodologies—particularly when it comes to different levels of technology maturity and the selection of specific indicators—presents a challenge for making objective comparisons across the CO_2 utilization sector [2]. For instance, the industry has yet to establish a standardized LCA protocol based on the International Organization for Standardization (ISO) guidelines for carbon capture and utilization (CCU). As a result, consistent

Chemical Engineering Department, Center for Carbon Capture, Storage and Utilization (CCUS), Institute of Sustainable Energy, Universiti Teknologi PETRONAS, Bandar Seri Iskandar, Perak, Malaysia.

* Corresponding author: anipeddi_22009697@utp.edu.my

and reliable "apples-to-apples" comparisons between technologies are difficult to achieve. The majority of CO_2 utilization technologies are in their infancy, with only a handful advancing to the demonstration plant stage, although many are projected to reach this phase. Advancing to the demonstration stage necessitates substantial investment, underscoring the necessity for transparent and rational assessment methods to guide funding allocation.

Carbon capture and utilization (CCU) involves the capture of CO_2 from sources such as industrial emissions or directly from the atmosphere, followed by its transformation into value-added products. CCU has demonstrated its capability to diminish environmental impacts, for instance, by reducing greenhouse gas emissions and lessening dependence on fossil fuels when compared to conventional technologies [3]. Nonetheless, the contribution of CCU to climate change mitigation is limited by the relatively small quantity of CO_2 that can be converted into products compared to the vast amounts currently being emitted. Moreover, many products derived from CO_2 are energetically unfavourable, meaning their formation usually requires significant energy input for chemical reduction. Conversely, CCU processes like mineralization do not require energy for the CO_2 conversion itself, but the preparatory stages, such as mineral grinding, can be energy-intensive and often exhibit slow reaction kinetics. Thus, the environmental and economic feasibility of CCU technologies frequently depends on specific contextual factors, including the availability of low-carbon electricity and economic conditions.

The last ten years have witnessed a remarkable surge in carbon capture and utilization (CCU) interest, evidenced by a significant increase in related scientific research. Numerous applied research institutions, established corporations, and emerging start-ups globally are actively developing CCU technologies, with several products already on the market such as CRI's Vulcanol, Covestro's Cardyon, and Carbon8's C8Agg [4]. A market study conducted in 2016 anticipated that CCU products could generate annual revenues reaching up to USD 800 billion by 2030, with the potential to utilize up to 7 billion tons of CO_2 annually.

Markets for Carbon Capture and Utilization (CCU) can be divided into two main categories: niche sectors and bulk markets. Niche sectors, such as plastics, chemicals, and carbon fibres, involve smaller quantities but yield higher profit margins. On the other hand, bulk markets deal in larger quantities but have lower profit margins and include industries like concrete, asphalt, and fuels. The substantial volume involved in these bulk markets presents a considerable opportunity for reducing emissions.

The burgeoning interest and optimistic forecasts for CCU are rooted in its several perceived economic and environmental benefits. CCU could serve as a cost-effective source of carbon, possibly replacing more costly alternatives. It presents innovative avenues for the synthesis of both existing and novel products, potentially carving out new markets. Furthermore, CCU is capable of addressing various challenges related to chemicals, fuels, materials, waste management, and the reduction of industrial CO_2 emissions. It provides a strategy for incorporating renewable energy into the chemical and transportation industries, promoting industrial synergy and a circular economy approach [5, 6]. CCU has the potential to simplify complex chemical reaction processes, improve overall efficiency, stabilize input costs, and lessen environmental impacts beyond just climate change mitigation. This is evident in CO_2-based fuels that can decrease nitrous oxide and soot emissions. When combined with CO_2 sequestration techniques like mineralization, CCU technologies have the potential to attain a carbon-negative status.

However, CCU is not without its challenges. The energy requirements for the majority of CCU processes are substantial, necessitating high-energy co-reactants, which could inflate both operational costs and environmental detriments. The development of new facilities, many of which operate at high pressures, requires significant capital investment. CCU is predominantly aimed at low-margin, high-volume industrial markets, which demand significant financial commitments. These industries, which include chemicals, fuels, and materials, often incur hefty costs in modifying existing processes and are characterized by a sluggish rate of product adoption [5, 7]. The reduction of environmental impacts stands as a pivotal criterion for the commercial success of CCU technologies; any CCU approach that fails to decrease overall environmental impacts is unlikely to find commercial success as an emissions mitigation strategy. Therefore, in light of the critical importance of both economic and environmental outcomes, it is imperative to conduct comprehensive evaluations. These assessments are typically performed using life cycle assessment for environmental analysis and techno-economic assessment for evaluating technical and economic feasibility.

5.2 Optimizing CO_2 Capture Strategies: A Comprehensive Approach through Techno-Economic Assessment and Environmental Analysis

TEA serves as a systematic framework for the evaluation of the technical prowess and economic feasibility of a process, product, or service. It delves into the economic ramifications of technology development stages including research, development, demonstration, and deployment, scrutinizing production costs and market potential. TEAs are often producer-centric, emphasizing the production phase, but they can be expanded to encompass life cycle stages, such as usage or end-of-life scenarios, to assess technical or economic viability across a product's life span.

It is critical to distinguish between assessment and decision-making, as TEA outputs inform subsequent technological enhancements or business strategy development. They are indispensable in various sectors for decision-making support, encompassing areas such as research, product development, investments, and policy-making [8]. TEA can facilitate decision-making for individual or multiple product lines and, within the context of this document, 'services' are also encapsulated under the term 'product.' While closely tied to technical development processes like chemical process design, TEA relies on data from process design and reciprocally influences design recommendations, yet does not encompass the technical development itself.

Conducted alongside ongoing research and development efforts, TEA aims to streamline the development process and expedite market entry. It is imperative to recognize that TEA is context-sensitive, relying on assumptions specific to factors like geographical location, time frame, and information accessibility. Although TEA addresses technical and economic questions, it deliberately omits environmental and social considerations.

TEA findings assist in making informed decisions tailored to specific projects, supporting R&D and investment choices. However, applying these findings to broader scenarios like global policy requires careful consideration due to their context-dependent nature.

In the domain of energy systems, a significant challenge lies in integrating a greater proportion of intermittent renewable electricity sources into the grid, which can cause substantial fluctuations in energy carrier prices. Consequently, processes are increasingly valued for their adaptability and ability to respond to power demands rather than just their output or efficiency [9]. Basic optimization models that primarily aim at reducing production costs might not be adequate in the evolving context. Additionally, regulatory elements, including carbon taxes and incentives for CO_2 capture, can significantly impact the financial dynamics of carbon capture approaches. While regulatory frameworks are evolving and vary across regions, stable regulations are crucial for making investment decisions and for the commercial viability of CO_2 capture technologies.

Concurrent with TEA, environmental and life cycle assessments (LCA) are vital for examining the climate-related impacts of processes. Defining the system boundaries and lifespan is fundamental for TEA and LCA analyses. While a technical lifespan of 15 to 20 years is standard for calculations, monitoring CO_2 storage may be necessary for much longer periods.

TEA of CO_2 capture processes frequently produce varied outcomes, mainly because of the variations in establishing study parameters, foundational cases, and the absence of extensive long-term data from large-scale implementations. Initiatives led by Rubin and colleagues have been critical in developing a standardized methodology for cost estimation in CO_2 capture, particularly for fossil fuel power plants, and later expanded to include a range of industrial processes. The scope of CO_2 capture technology now extends to various industries such as steel, cement, petrochemicals, and hydrogen production, as detailed in Yang et al.'s study [10, 11]. This diversification introduces greater complexity into the techno-economic evaluations compared to those for traditional thermal power plants.

Applying second law-based or exergy analyses has become pivotal in comparing and optimizing different CO_2 capture systems, especially in multi-product cogeneration plants. These analyses help in fine-tuning processes and ensuring sustainable resource utilization. Clear definition of system boundaries, adopting approaches like cradle-to-grave and gate-to-gate, is essential in these studies. The focus is on key aspects such as technical performance, cost structures, and CO_2 mitigation potential, while also briefly touching on Life Cycle Assessment methodologies. This highlights the importance of adaptable and uniform methods in assessing the environmental and economic impacts of CO_2 capture technologies across various industrial sectors.

5.3 Key Economic Parameters and Cost Analysis for CO_2 Capture Plants

To accurately estimate the costs associated with CO_2 capture, it is essential to define realistic parameters for each case. The primary factors influencing these costs include:

Plant Full Load Hours (h/a): This term refers to the yearly operational hours of a plant, a vital consideration in assessing the economic viability of CO_2 capture. Plants involved in industrial activities and baseline energy production, especially those suitable for CO_2 capture, generally function for about 7,000 to 8,000 hours each year.

Operational Lifetime (years): In the energy and chemical processing sectors, the lifespan of plants generally ranges from 15 to 25 years.

Interest and Inflation Rates (%): The actual cost of money is calculated by the difference between these two rates. Recently, these rates have been relatively low, with interest rates usually falling between 3 and 8%.

Time for Construction (years): The duration needed for the planning, obtaining permits, and building of a new plant greatly influences its total cost. This phase can extend over several years, particularly for large-scale facilities.

CO_2 capture plant expenses are generally divided into two main categories: Capital Expenditures (CAPEX) and Operational Expenditures (OPEX). CAPEX encompasses all initial investments, such as land purchase, materials procurement, planning, engineering, and construction. These

costs are further divided into Bare Erected Costs (BEC), which cover the costs of process equipment, labour, Engineering Procurement, and Construction (EPC), along with contingency costs for both the process and the project. The Total Plant Costs (TPC) include all these aspects.

In addition to TPC, major projects typically have Owner's Costs, which account for feasibility studies, land acquisition, insurance, permits, and financing. When added to TPC, these constitute the Total Overnight Cost. A thorough analysis of the BEC components is critical, with tools like the Aspen Process Economic Analyzer being useful for assessing CAPEX and OPEX of standard chemical process elements. EPC costs are often calculated as a percentage of BEC, and contingency costs fluctuate based on the project's level of development [12, 13].

OPEX divides into variable and fixed costs. Variable costs include energy, consumables, and costs associated with CO_2 capture, treatment, and disposal processes. For comprehensive CCS chains, these also extend to CO_2 transport and storage. Fixed OPEX, often a percentage of variable costs, includes personnel, maintenance, taxes, and insurance. The operation of CO_2 capture plants is usually managed by existing personnel, with additional staffing costs not significantly affecting the overall operational expenses. However, due to limited data from large-scale CO_2 capture plants, these costs are often estimates based on the broader chemical and process industry.

The cost of CO_2 captured, expressed in dollars per ton of CO_2 ($/t$CO_2$), is calculated using the following formula [22]:

The cost per ton of CO_2 captured can be calculated using the following formula:

$$\text{Cost of CO}_2\text{ Captured}(\$/\text{tCO}_2) = \frac{\text{Cost of Electricity with Capture} - \text{Cost of Electricity for the Reference Plant}}{\text{Total CO}_2\text{ Captured per Net MWh for the Plant with Capture Cost of CO}_2\text{ Captured}}$$

In this formula:

"Cost of Electricity with Capture" refers to the electricity cost for the plant that includes CO_2 capture technology.

"Cost of Electricity for the Reference Plant" represents the cost of electricity for a similar plant but without CO_2 capture technology.

"Total CO_2 Captured per Net MWh for the Plant with Capture" is the net amount of CO_2 captured by the plant, calculated as the total CO_2 produced minus the CO_2 emitted.

It is important to note that the cost of CO_2 captured here excludes expenses related to the transport and storage of CO_2. This metric specifically measures the cost associated with the capture (production) of CO_2 only.

The Cost Of Electricity generation (COE) in power plants, particularly when considering carbon capture and storage (CCS) technologies, is influenced by a variety of factors. This cost can be calculated using the following formula [22]:

$$COE = \frac{TCC \times FCF + FOM}{CF \times 8766 \times MW} + VOM + HR \times FC$$

Where:

COE represents the cost of generating electricity, measured in \$/MWh.

TCC is the total capital cost of the plant, in dollars.

FCF stands for the fixed charge factor, which is a yearly fraction.

FOM denotes fixed operating and maintenance costs, in dollars per year.

VOM refers to variable non-fuel operating and maintenance costs, in \$/MWh.

HR is the net power plant heat rate, in MJ/MWh.

FC indicates the unit fuel cost, in \$/MJ.

CF is the plant capacity factor, a fraction representing the ratio of actual output to maximum possible output.

8766 is the total number of hours in an average year.

MW signifies the net plant capacity in megawatts.

This formula indicates that the COE (Cost of Electricity) is dynamic and influenced by various elements, including the capacity factor of the plant, the cost of fuel, and other operational expenses. Given that these factors can vary throughout a power plant's operational life, the COE is subject to annual fluctuations. In studies examining the engineering and economic dimensions of CCS (Carbon Capture and Storage) at the level of individual power plants, the total cost of CCS implementation is assessed by comparing the COE of a plant with CCS to that of a comparable plant without CCS technology. It is essential to clearly understand the specific scenario, whether it concerns single plants or a collective of plants, to accurately analyze and interpret these costs.

The concept of CO_2 avoidance cost is a critical metric in evaluating the economic feasibility of CO_2 capture and storage (CCS) technologies, especially for power plants. This cost measure, referred to as CCO_2 avoided, is derived by comparing the Levelized Cost Of Electricity (LCOE) between a plant equipped with CCS and a reference plant without CCS. The key formula for calculating the cost of avoided CO_2 is [15, 22]:

$$CO_2\ avoidance\ cost = \frac{LCOECCS - LCOEref}{EFref - EFCCS}$$

In this context, LCOE_CCS and LCOE_ref denote the levelized costs of electricity generation for plants equipped with CCS and for standard reference plants without CCS, respectively.

EF_ref and EF_CCS represent their respective emission factors, measured in tons of CO_2 per megawatt-hour of electricity produced.

It is crucial to understand that plants with CCS are not completely devoid of emissions, as the process of capturing CO_2 itself consumes energy and results in emissions. Thus, the effective amount of CO_2 that is avoided is invariably lower than the total quantity captured. This disparity leads to a higher cost per unit of CO_2 avoided as opposed to the cost of capturing CO_2. The efficiency with which CO_2 is captured significantly influences these cost dynamics. Instead of pursuing 100% capture efficiency, which becomes increasingly expensive for the final portions of CO_2, a more feasible approach is to optimize the balance between capture efficiency and the cost of CO_2 avoidance [14, 15].

This consideration is especially relevant for CO_2 capture from fossil fuel sources. For fossil-based emissions, any capture efficiency less than 100% still results in a net increase of atmospheric CO_2. In contrast, capturing CO_2 from CO_2-neutral sources like sustainable biomass or direct air capture can reduce atmospheric CO_2 levels, even with less than 100% efficiency [16].

In industrial settings or cogeneration plants producing multiple outputs, defining CO_2 avoidance costs requires a different approach. It is crucial that both the reference and CCS-equipped plants produce the same output, whether it is steel, cement, pulp, or biofuels. The formula for these settings would be adjusted to account for the levelized cost per unit of key products and their associated emission factors [26, 27, 28].

In cogeneration plants, CO_2 capture can alter the optimal balance of different products, such as changes in the amount of heat produced or used. One way to compare different outputs is through exergy-based analysis, converting electricity and heat into exergy flows and evaluating CO_2 avoided costs based on the levelized cost of exergy.

Moreover, when allocating costs in cogeneration plants, a method involving dividing costs into joint and separable costs can be applied. This approach requires careful consideration of the basis for cost allocation, whether by mass, energy content, market value, or assuming zero value for by-products.

These intricacies highlight the challenges in standardizing and accurately determining CCS costs, as noted in the document "Understanding the pitfalls of CCS cost estimates" [1]. The differences in methodologies, assumptions, and the complexity of comparing various outputs underscore the need for a thorough and nuanced approach to evaluating CCS technologies.

5.4 Methodological Steps for Techno-Economic Analysis of CO_2 Capture Processes

5.4.1 Overview of Techno-Economic Analysis (TEA) Procedure

Techno-Economic Analysis (TEA) is a systematic approach used to evaluate the feasibility of CO_2 capture processes, both existing and new technologies. The depth of these studies varies based on the project's stage—from preliminary scans of new technologies to detailed analyses for imminent plant constructions. The key steps in most TEA studies include:

5.4.2 Characterizing the Case Study

The characterization involves identifying the type and capacity of the plant, determining the products and production volumes, assessing the needed fuels and raw materials, and considering location-specific regulations and conditions.

Process Modelling Approach: This encompasses gathering data from existing plants, literature, or creating models for plants without CO_2 capture, selecting CO_2 capture methods for the study, conducting detailed process modelling for the CO_2 capture plant, which includes calculating energy and mass flow rates and technical performance indicators, and listing the necessary components for the process.

Cost Analysis Methodology: The methodology sets key economic parameters like the discount rate, plant lifetime, operational schedule, and load profile. It estimates prices for products, raw materials, fuels, and utilities, including taxes and emission prices in the calculations, and calculates the capital and operational costs of the plant. This also involves determining economic indicators such as LCOE and the Cost of CO_2 avoided.

Environmental Impact Evaluation: This phase measures plant emissions and compares CO_2 emission reductions to a reference case, conducts a Life Cycle Analysis (LCA), and focuses on determining the plant's energy and mass flows. Detailed process modelling, essential for full cost calculations, usually utilizes process simulation software, and economic calculations should consider price variations and conduct sensitivity analysis for key parameters.

Techno-Economic Analysis (TEA) Optimization and Sensitivity Examination: TEA often incorporates techno-economic (environmental) optimization, usually an iterative process, particularly complex in multi-product plants. Methods like Monte Carlo simulations, testing numerous scenarios with variable main parameters, are used for optimization. This, combined with sensitivity analysis, provides insights into plant performance and cost variability.

Comparing Costs Across Varied CO_2 Capture Technologies: Assessing costs in different CO_2 capture technologies involves understanding how cost structures vary and are critical in determining the feasibility of these technologies in different operational environments. Rubin et al. proposed a standardized methodology for presenting CO_2 capture plant costs, with reference values for various technologies suggesting average values from diverse sources [19, 20].

5.5 Key Considerations in Cost Structure Analysis

- Identifying the critical cost components for each technology.
- Facilitating technology comparisons and initial sensitivity analysis.
- Considering variables like electricity, heat, and construction material costs.

A case study was conducted on a 200 MWth biomass boiler with 12 vol-% CO_2 content at the outlet. The assumptions included 7500 full load hours per year and set costs for electricity and heat. This comparison showed that technologies like MEA require significant regeneration heat, whereas processes like pressure swing adsorption need more electricity for flue gas compression. Technology Readiness Level (TRL) also impacts process and project contingency costs, affecting the overall cost structure.

5.6 Prospects for Cost Reduction in CO_2 Capture Technologies

Limitations and Potential of Emerging Technologies

While emerging CO_2 capture technologies may offer improvements in capacity and durability, significant cost reductions on a large scale are unlikely. For instance, advancements in sorbent materials and efficiencies achieved through mass production could lower costs, but drastic reductions are not anticipated. Instead, the primary source of cost reduction is expected to come from scaling up current pilot-scale facilities and applying lessons learned from existing installations as the number of full-scale projects increases.

The Role of Learning Curves and Scaling in Cost Reduction

The scarcity of cost information from large-scale, commercially operating CO_2 capture facilities underscores that the initial expenses of first demonstration plants are generally greater than those of later installations.

To comprehend how these costs change over time, terms such as "First of a Kind" (FOAK) and "Nth a The Kind" (NOAK) are employed. Rubin developed a method for evaluating learning curves in different energy technologies, which is commonly expressed as a power function [28].

$$y = ax^{-b}$$

Here, 'y' represents the capital cost for the nth unit per unit capacity, 'a' is the initial unit's capital cost per capacity, 'x' is the cumulative capacity ratio, and 'b' is the learning rate exponent. The Learning Rate (LR), which indicates the percentage reduction in cost with each doubling of capacity, can be expressed as [28]:

$$b = \frac{\log(1 - LR)}{\log 2}$$

Furthermore, cost reduction through scaling up pilot and demonstration plants plays a significant role. Costs for various unit sizes are often evaluated using an exponential scale-up law, characterized by a scaling factor (sf) [28, 29]:

$$C = Cref\left(\frac{S}{Sref}\right)^{sf}$$

In this context, 'C' represents the capital cost of a component, and 'S' refers to its size parameter, with 'ref' indicating reference values. The size parameter is generally associated with the component's capacity, like the fuel power for boilers or the mass flow rates for pipelines. The scaling factors, which typically range from 0.5 to 1, differ among various components of energy systems. For example, reactor vessels and boilers usually apply a scaling factor of 0.67, unless more specific information is available. In cases of cost components where scale economies do not play a role, such as in variable operating and fuel costs, the scaling factor is assumed to be 1.

5.7 Challenges and Future Perspectives in Techno-Economic Analysis of CO_2 Capture

Current Challenges in TEA of CO_2 Capture Processes

Limited Long-term Operational Data: A primary challenge in conducting TEA for CO_2 capture processes is the scarcity of long-term operational data from large-scale facilities. This gap necessitates reliance on data extrapolated from smaller-scale pilot or demonstration plants, or analogous processes, introducing uncertainties in cost estimations.

Uncertainty in Future Cost Variables: Predicting future prices for fuels, electricity, and other consumables is fraught with uncertainty. This is especially pertinent given the increasing integration of variable renewable energy sources into the grid, affecting electricity pricing dynamics.

Need for Flexibility in Capture Processes: The evolving energy landscape, characterized by fluctuating availability and pricing of electricity, demands greater flexibility in CO_2 capture processes [21]. This variability necessitates more sophisticated dynamic simulations over extended periods to accurately represent part-load behaviour and operational flexibility.

Infrastructure and Legislative Hurdles for CCUS Technologies

Infrastructure Development: The development of comprehensive infrastructure for CO_2 transport, storage, and utilization remains a significant hurdle. Substantial investment is required to establish efficient and secure CO_2 handling procedures.

Legislative Uncertainty: A lack of uniform legislation governing CO_2 transport and storage contributes to uncertainties, delaying investment decisions in urgently needed CO_2 capture technologies [22]. The costs associated with these Off-Site Balance of Plant (OSBL) aspects pose substantial challenges in techno-economic analysis and lifecycle assessments.

Future Market Potential and Technological Readiness

CO_2 Economy Opportunities: Despite these challenges, the emerging CO_2 economy presents significant market potential, comparable to the current oil and gas industry. This potential stems from the growing need for carbon removal technologies and the shift away from fossil-based carbon sources.

Technological Viability and Economic Drivers: Various CO_2 capture technologies are available for deployment at an industrial scale, provided economic conditions are favourable. The estimated cost range for capturing CO_2 from industrial sources is between Euros 20 to 100 per ton of CO_2.

Rising CO_2 Prices as a Business Incentive: The increasing price of CO_2, as observed in systems like the European Emission Trading Scheme, where prices have exceeded Euros 80 per ton (as of December 2021), is beginning to make CO_2 capture projects economically viable. These rising prices could create profitable business cases for CO_2 capture initiatives, accelerating their adoption and implementation.

In conclusion, while TEA of CO_2 capture processes faces significant challenges, particularly in data availability and infrastructure development, the evolving energy market and legislative landscape offer promising opportunities [24, 25]. The increasing valuation of CO_2 emissions and the market potential for carbon removal technologies suggest a promising future for CO_2 capture initiative.

References

[1] Rubin, E. S., Short, C., Booras, G., Davison, J., Ekstrom, C., Matuszewski, M. et al. 2013. A proposed methodology for CO_2 capture and storage cost estimates. Int. J. Greenh Gas Control 2013: 488e503.
[2] Roussanaly, S., Rubin, E. S., van der Spek, M., Booras, G., Berghout, N., Fout, T. et al. 2021. Towards improved guidelines for cost evaluation of carbon capture and storage (Version 1.0). 29 3. 2021. https://doi.org/10.5281/zenodo.4643649 [Accessed 20 July 2021].
[3] Yang, F., Meermana, J. C. and Faaij, A. P. C. 2021. Carbon capture and biomass in industry: a technoeconomic analysis and comparison of negative emission options. Renew. Sust. Energ Rev. 2021: 144.
[4] Lara, Y., Martı´nez, A., Lisbona, P., Bolea, I., Gonza´lez, A. and Romeo, L. M. 2011. Using the second law of thermodynamic to improve CO_2 capture systems. Energy Proc. 2011: 1043e50.
[5] Ziebik, A. and Gładysz, P. 2015. Thermoecological analysis of an oxy-fuel combustion power plant integrated with a CO_2 processing unit. Energy 2015: 37e45.
[6] Peltola, P., Tynja¨la¨, T., Ritvanen, J. and Hyppa¨nen, T. 2014. Mass, energy, and exergy balance analysis of chemical looping with oxygen uncoupling (CLOU) process. Energy Convers Manag 2014: 483e94.
[7] Saari, J., Peltola, P., Tynja¨la¨, T., Hyppa¨nen, T., Kaikko, J. and Vakkilainen, E. 2020. High-efficiency bioenergy carbon capture integrating chemical looping combustion with oxygen uncoupling and a large cogeneration plant. Energies 2020: 3075.
[8] Atsonios, K., Panopoulos, K., Grammelis, P. and Kakaras, E. 2016. Exergetic comparison of CO_2 capture techniques from solid fossil fuel power plants. Int. J. Greenh Gas Control 2016: 106e17.
[9] De Lena, E., Spinelli, M., Gatti, M., Scaccabarozzi, R., Campanari, S., Consonni, S. et al. 2019. Techno-economic analysis of calcium looping processes for low CO_2 emission cement plants. Int. J. Greenh Gas Control 2019: 244e60.
[10] EC. "EUR-Lex." Commission Delegated Regulation (EU) 2015/2402 of 12 October 2015 reviewing harmonised efficiency reference values for separate production of electricity and heat in application of Directive 2012/27/EU of the European Parliament and of the Council and rep. 2015. http://data.europa.eu/eli/reg_del/2015/2402/oj.
[11] EEA. Greenhouse gas emission intensity of electricity generation in Europe. 2021. https://www.eea.europa.eu/data-and-maps/indicators/overview-of-the-electricity-production-3/assessment-1.
[12] Volker, Q. 2021. Specific carbon dioxide emissions of various fuels. 5. Haettu, https://www.volker-quaschning.de/datserv/CO2-spez/index_e.php.
[13] Dittmann, A., Sander, T. and Robbi, S. 2008. Allocation of CO_2-emissions to power and heat from CHPplants. Dresden: TU Dresden; 2008. https://tu-dresden.de/ing/maschinenwesen/iet/gewv/ressourcen/dateien/veroefftlg/alloc_co2.
[14] Peltola, P., Saari, J., Tynja¨la¨, T. and Hyppa¨nen, T. 2020. Process integration of chemical looping combustion with oxygen uncoupling in a biomass-fired combined heat and power plant. Energy 2020: 118550.

[15] Rubin, E., Booras, G., Davison, J., Ekstrom, C., McCoy, S., Short, C. et al. 2021. Toward a common method of cost estimation for CO_2 capture and storage at fossil fuel power plants. A white paper prepared by Task Force on CCS Costing Methods. 2013 [Accessed 20 July 2021], https://www.cmu.edu/epp/iecm/rubin/PDF%20files/2012/CCS%20Task%20Force_White%20Paper_FINAL_Jan%2015%202013.pdf.
[16] Hyva¨rinen, J. 2019. Techno-economic evaluation of carbon capture technologies integrated to flexible renewable energy system. Lappeenranta [Masters's thesis]. LUT University.
[17] Aspen Technology Inc. Aspen process economic analyzer. 2021 [Accessed 23 July 2021], https://www.aspentech.com/en/products/pages/aspen-process-economic-analyzer.
[18] Gardarsdottir, S. O., De Lena, E., Romano, M., Roussanaly, S., Voldsund, M., Pe´rez-Calvo, J-F. et al. 2019. Comparison of technologies for CO_2 capture from cement production. part 2: cost analysis. Energies 542: 1e20.
[19] Li, K., Wardhaugh, L., Paul, F., Yu, H. and Moses, T. 2016. Systematic study of aqueous monoethanolamine (MEA)-based CO2 capture process: techno-economic assessment of the MEA process and its improvements. Appl. Energy 2016: 648e59.
[20] Van der Spek, M., Eldrup, N. H., Skagestad, R. and Ramı´rez, A. 2017. Techno-economic performance of state-of-the-art oxyfuel technology for low-CO_2 coal-fired electricity production. Energy Proc. 2017: 6432e9.
[21] Lin, Z., He, Y., Li, L. and Wu, P. 2018. Tech-economic assessment of second-generation CCS: chemical looping combustion. Energy 2018: 915e27.
[22] Rubin, E. S. 2012. Understanding the pitfalls of CCS cost estimates. Int. J. Greenh Gas Control 2012: 181e90.
[23] Deevski, S. 2016. Cost allocation methods for joint products and by-products. Econ Altern 2016: 64e70.
[24] Gorre, J., Ruoss, F., Karjunen, H., Schaffert, J. and Tynja¨la¨, T. 2020. Cost benefits of optimising hydrogen storage and methanation capacities for Power-to-Gas plants in dynamic operation. Appl. Energy 2020: 11367.
[25] DOE/NETL. 2013. Cost and performance baseline for fossil energy plants volume 1: bituminous coal and natural gas to electricity. DOE/NETL. https://www.netl.doe.gov/sites/default/files/netl-file/BitBase_FinRep_Rev2.pdf.
[26] Wunderlich, J., Armstrong, K., Buchner, G. A., Peter, S. and Reinhard, S. 2021. Integration of technoeconomic and life cycle assessment: defining and applying integration types for chemical technology development. J. Clean Prod. 2021: 125021.
[27] Terlouw, T., Bauer, C., Rosa, L. and Mazzotti, M. 2021. Life cycle assessment of carbon dioxide removal technologies: a critical review. Energy Environ Sci. 2021: 1701e21.
[27a] Rubin, E. S. 2019. Improving cost estimates for advanced low-carbon power plants. Int. J. Greenh Gas Control 88: 1e9.
[28] Guandalini, G., Romano, M. C., Ho, M., Wiley, D., Rubin, E. S. and Carlos Abanades, J. 2019. A sequential approach for the economic evaluation of new CO_2 capture technologies for power plants. Int. J. Greenh Gas Control 2019: 219e31.
[29] Turner, M. J. and Pinkerton, L. L. 2013. Quality guidelines for energy system studies: capital cost scaling methodology. DOE. https://doi.org/10.2172/1513277.

Chapter 6

CO_2 Sequestration via Clathrate Gas Hydrates

Abdirahman Hassan Mohamed,[1,*] *Aliyu Adebayo Sulaimon*[1] and *Bhajan Lal*[2]

6.1 Introduction to Geological Storage

6.1.1 CO_2 Trapping Mechanisms

Once CO_2 is captured and transported, it is injected into a subsurface marine or terrestrial geologic sediment for storage. The injected CO_2 fluid will diffuse latterly and upwards from the injection point, the CO_2 fluid flow is influenced by capillary, viscous, and gravity forces in the porous medium. Eventually this mobile CO_2 phase will be demobilized by different trapping mechanisms that also determine the storage capacity of the storage sediment [1]. The main CO_2 trapping mechanisms are structural trapping, solubility trapping, residual trapping, and mineralization of the CO_2 phase within the host sediment [2, 3]. In order to maximize the storage capacity and security of the CO_2 fluid, it is injected as a supercritical fluid, hence above the critical pressure and temperature of the CO_2 (Pc = 7.38 MPa; Tc = 304 K). Therefore, this necessitates injecting the CO_2 at depths where the hydrostatic pressure exceeds the CO_2 critical pressure and temperature, which is typically encountered at approximately 800 m deep. Supercritical CO_2 (sc-CO_2) exhibit lower viscosity, higher diffusivity, and similar density compared to a liquid CO_2 phase; and it exhibits lower

[1] Department of Petroluem Engineering, Universiti Teknologi PETRONAS, Bandar Seri Iskandar, Perak Darul Ridzuan, Malaysia.

[2] Chemical Engineering Department, Center for Carbon Capture, Storage and Utilization (CCUS), Institute of Sustainable Energy, Universiti Teknologi PETRONAS, Bandar Seri Iskandar, Perak, Malaysia.

* Corresponding author: Mohamed_25550@utp.edu.my

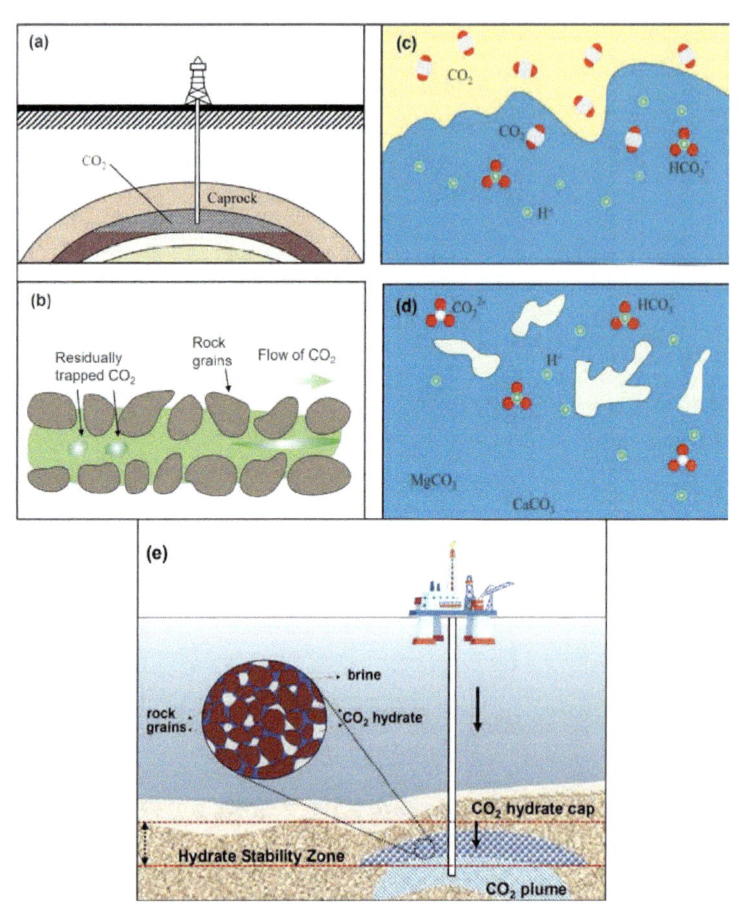

Figure 1. CO_2 Trapping Mechanisms—(a) structural trapping; (b) Residual trapping; (c) Solubility trapping; (d) Mineral trapping; (e) Hydrate formation (Modified with permission from [5]).

mobility compared to a CO_2 gas. Hence, due to these properties the sc-CO_2 is utilized in many applications such as Enhance Oil and Recover (EOR); shale gas drilling and fracturing; and CO_2 storage [4].

Structural trapping occurs when the CO_2 is physically trapped by an impermeable caprock in the storage sediment. Typically, when CO_2 is injected into deep saline aquifers and depleted oil and gas reservoirs (Figure 1a), the CO_2 migrates upwards and the presence of caprocks such as shale layers in the reservoir will prevent further vertical diffusion of the CO_2 and leakage. This trapping mechanism traps the greatest amounts of CO_2 globally. Moreover, at low temperature, high pressure, and shallow marine sediment depths lies the Hydrate Stability Zone (HSZ). Hence, as the CO_2 diffuses vertically it will reach the hydrate zone, whereby it forms layer of impermeable **hydrate capping** layer that will also prevent CO_2 leakage (Figure 1e).

Residual trapping occurs when the injected CO_2 saturation increases in the sediment pores, the water saturation reduces simultaneously, then as the buoyant CO_2 migrates upwards, the water reoccupies those pores as the CO_2 saturation decreases, hence there will be an amount of CO_2 that becomes immobile, and is called residually trapped CO_2 or irreducible CO_2 saturation (Figure 1b).

Solubility trapping is a long-term process, whereby a portion of the injected CO_2 will gradually dissolve into the formation brine. Solubility of the CO_2 depends on reservoir pressure and temperature, geochemistry, mineralogy, and salinity of the water in the storage sediment (Figure 1c).

Mineral trapping is the slowest and most secure CO_2 trapping mechanism. Mineralization of CO_2 is a geochemical reaction, whereby the dissolved CO_2 reacts with minerals of the rock matrix, which results in the carbonate precipitation and formation of carbonate minerals within the pore space (Figure 1d).

6.1.2 CO_2 Mineralization

Mineral carbonation refers to the injection of captured CO_2 into underground reactive rock minerals such as mafic or ultramafic rocks, in order to achieve CO_2 mineralization into stable carbonate forms such as magnesite ($MgCO_3$), siderite ($FeCO_3$), and calcite ($CaCO_3$), within the formation or in an artificial process [6, 7]. CO_2 mineralization is a potential storage pathway which is permanent. This reaction exists in nature, for instance silicate weathering, which happens in long geological timescales. Hence, artificial CO_2 mineralization can significantly reduce the required time for CO_2 conversion to carbonate minerals from carbonated water [7].

Interestingly, within two years of injecting carbonated water into these reactive rocks and minerals, at temperature ranges (20–50°C), most of the CO_2 is trapped and converted into carbonate minerals [8, 9]. Moreover, the injectant CO_2 can dissolve in water before or while injecting. Carbonated water (CO_2 + water) is denser than formation water, hence CO_2 does not diffuse upwards, and the sealing formation is not necessary. Currently, carbon mass in the atmosphere is estimated at 800 Gt. On the contrary, carbonate rocks such as limestone and marble, hold a significant amount of carbon mass that is estimated at 39 million Gt. Therefore, this indicates the possibility of storing CO_2 within underground formation for mineralization [10, 11]. The *in-situ* mineralization process involves CO_2 mixing with water, which forms acidic water with pH values ranging (3–5) before it is injected to the formation [12, 13]. The acidic condition will cause silicate mineral dissolution. Eventually, carbonate minerals start to precipitate as the acidic carbonated water neutralizes. Moreover, cations can convert to carbonate minerals with carbonated water. Usually, calcium ions form calcite, while magnesium ions form magnesite and dolomite [12].

6.1.3 CO_2 Hydrate Formation

At high pressure, low temperature CO_2 hydrate starts to form in the presence of CO_2 gas/liquid and water. The narrow pressure and temperature conditions at which CO_2 hydrates form are known as phase equilibrium conditions. Moreover, the marine sediment and permafrost location provide suitable conditions for hydrate formation. Hydrate-bearing sediments in these locations are characterized with low permeability. In 1995, Koide et al. [14] proposed an injection of CO_2 into underground sediments, to form CO_2 hydrates as means of permanent storage. Injected CO_2 will migrate upwards due to its buoyancy (density difference between formation brine and the CO_2 phase). When the CO_2 reaches the hydrate formation zone, CO_2 hydrate starts to form under high-pressure and low temperature conditions [15]. Consequently, the CO_2 hydrate forms a capping layer that prevents further vertical migration of CO_2 (see Figure 1e). This can be an extra layer of sealing and security for sediments with stratigraphic impermeable caprock. Moreover, beneath marine sediments (more than 300 m depth), the sediments comprise alternating mud-sand-mud layers. After CO_2 gas is injected into these sediments CO_2 will migrate to sand layers to form hydrates and any possible leakage would be contained by the low-permeability mud layers.

6.2 Reservoir Rock-Fluid Interactions

6.2.1 Wettability, Capillary Pressure, and Relative Permeability

Wettability, capillary pressure, and relative permeability are the three characteristics of rock-fluid interactions that govern fluid flow processes of imbibition and drainage, which occur when CO_2 is injected into and propagates through a storage sediment [16]. The concepts of relative permeability and associated models are well-established in reservoir fluid production literature (e.g., [17, 18]). These models aim to predict how a fluid phase flows relative to the presence of a second fluid, considering the fluid phase spatial distribution within the porous medium. At pore scale, fluid phase distribution is governed by wettability (surface energies). In the presence of two fluid phases such as CO_2 and water, one phase will wet the rock surface, while the other phase will occupy and fill the open pores [19].

Figure 2 shows the relative permeability curves of gas and water in a Berea sandstone. It further illustrates the simplified fluid phase distribution

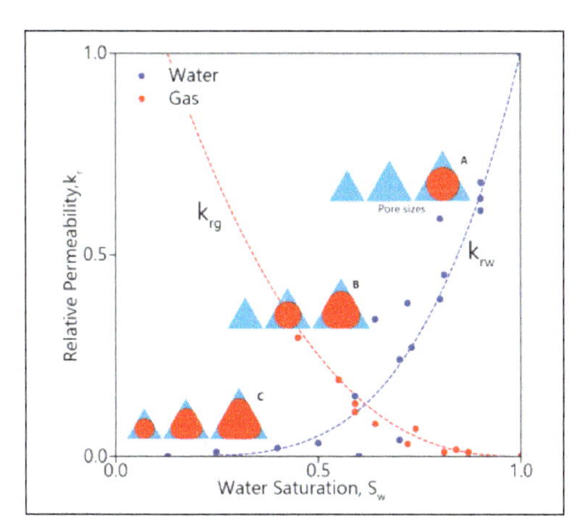

Figure 2. Relative Permeability curves of gas/water system, and fluid phase pore occupancy (Reproduced with permission from [19]).

in different-sized triangular pore spaces. The blue curve represents the wetting phase (brine in this case), and the red curve is the non-wetting phase (gas in this case). The wetting fluid phase minimizes the grain/ mineral surface energy as it covers and spreads around the grain as narrow fluid film. Water (brine) is the wetting phase in most geologic materials, whereas oil or gas are typically the non-wetting phases. When wetting phase saturation is high, water occupies most pores, except the larger pores, which have the least interfacial curvature (Figure 2A). As the wetting-phase decreases, water is progressively pushed into smaller pores, thus increasing the interfacial curvature as depicted in (Figure 2B and C).

In the porous media, the largest pores are the best channels for fluid flow; Therefore, as the saturation of the wetting-phase decreases its relative permeability declines (Figure 2, blue curve, right to left). Simultaneously, the relative permeability of the non-wetting-phase increases with saturation (Figure 2, red curve, right to left). However, the rate of relative permeability declines with decreasing phase saturation is path dependent. In Figure 2 (red curve, left to right), relative permeability declines slower with the decreasing non-wetting phase, compared with the wetting phase. Furthermore, imbibition is a process that increases the wetting-phase saturation—such as waterflooding of water-wet reservoirs. Conversely, drainage is a process that reduces the wetting-phase saturation—such as injecting CO_2 into an aquifer (wetting brine saturation decreases). After the CO_2 injection ceases, the wetting fluid (brine water) flows back into the pores behind the rising buoyant CO_2 plume, this is an imbibition process [16].

Relative permeability curves for each phase are usually generated with traditional correlations such as [20, 21], which were developed for oil and gas reservoirs. Each porous medium has a distinctive geometry defined by geological factors such as a depositional environment and weathering cycles; hence fluid flow behaviour will vary for different sediments. Furthermore, relative permeability curves will also differ depending on the wettability of the reservoir rock and the direction of the fluid flow. For instance, in dynamic conditions contact angel (degree of wettability) changes with fluid flow path. Under drainage the contact angle reduces, indicating higher wettability (affinity to rock). Conversely, under imbibition, contact angle increases, exhibiting lower wettability. This behaviour affects capillary pressure and permeability within the porous media, and it is well known as hysteresis [19].

During the production from a gas hydrate reservoir, effective porosity which is the interconnected pore channel that allows fluid flow, is constantly changing due to the dissociation of gas hydrates. Therefore, using traditional models such as Brooks-Corey and van Genuchten for accurate permeability prediction would require adjustments with empirical parameters for the range of varying hydrate saturations. However, due to absence of experimental relative permeability data, such calibrations are challenging [22]. Despite the solid state, hydrates exhibit mobility through dissolution and reformation [23]. Regardless of the hydrate nucleation process in a porous medium, its growth from a nuclei mimics a mobile, non-wetting phase that will self-rearrange to minimize energy through the Ostwald ripening phenomenon [24, 25]. Moreover, according to Waite et al. [26], grain surfaces have lower interfacial energy when contacted by water than hydrate.

The non-wetting behaviour of hydrate growth in porous media concept was introduced by [27]. However, in context of relative permeability, capillary tube model of the porous media was mostly applied [28]. Hence, this simplification provides the prediction of the functional form of relative permeability of fluid in hydrate bearing sediments. Models that describe hydrate behaviour and morphology in porous media (i.e., wetting and grain coating; non-wetting and pore filling), have served as the guiding assumption for many studies that analyze sediment properties such as relative permeability, acoustic, electrical, and hydraulic properties [29–31]. However, these simplified models have not captured the actual physical complexity of hydrate-bearing sediments.

In gas hydrate-bearing sediments, the hydrate morphology influences the fluid phase flow behaviour. Murphy et al. [19] performed steady-state relative permeability experiments in Berea sandstone core samples with different hydrate saturations. They measured and compared relative permeability of water in the presence of gas and in the presence of hydrate. The results indicated that the water relative permeability in the hydrate/water and gas/water systems behaved similarly. Fluid flow behaviour is highly dependent on porous media properties, which have different pore geometry influenced by geologic factors such as parent material, depositional environment, diagenesis, and weathering ([17]); hence the shape of the relative permeability curve will vary. Their study concluded that (A) hydrate forms as the non-wetting phase, preferentially filling the large pores; (B) Since the relative permeability in both systems were the same, relative permeability in hydrate/water systems can be accurately derived through simpler gas/water relative permeability measurements in the same hydrate-bearing sediment; (C) models that assume a fixed pore or tube geometry are inadequate, and relative permeability is the simplest framework to understand hydrate-induced permeability changes.

[22] proposed a new relative permeability model for multiphase flow in gas hydrate-bearing sediments. The three unique features about their model are (A) experimental data for one hydrate saturation are required to fit 6 model parameters once, subsequently these parameters could be used to predict relative permeability at any hydrate saturation; (B) it includes effect of capillarity; (C) accounts for the effect of pore-size distribution. Hence, these features are and not accounted for in currently used numerical simulators for hydrate-bearing reservoirs. Therefore, to achieve accurate prediction of fluid relative permeability in hydrate reservoirs, Brooks-Corey, and van Genuchten, and other legacy models would require their parameter to be a function of hydrate saturation, to account for its effect on changing porosity and permeability due to dissociating hydrates during production. However, that would be a resource intensive and time-consuming process because it requires experimental data of relative permeability over many hydrate saturations. Furthermore, numerical simulations were performed comparing their proposed model and Brooks-Corey (B-C) model [22]. They concluded that both models took the same time to run simulations on two studies reservoirs. Predictions employing their proposed model differed from the B-C model. Hence, that showed the inherent approximation of the B-C model when applied across a range of hydrate saturations, when contrasted with the proposed model which incorporates change in hydrate saturation, for accurate prediction of reservoir performance [22].

6.3 Conventional Geological CO_2 Storage Pathways

6.3.1 Introduction to Geological Storage

Over the years increasing emissions of greenhouse gases, mainly CO_2 from various industrial activities and fossil fuel combustion has hastened global warming which adversely impacted weather patterns and ecological systems. Therefore, anthropogenic CO_2 is captured from combustion sources, it is then transported, utilized, and stored for permanent containment. Currently geological formations such as water-bearing formations, oil and gas reservoirs are technically ready and suitable storage options for large-scale CO_2 injection and containment. Aspects such as: storage sediment characterization, CO_2 injection and monitoring technologies closely resemble those utilized for oil and gas production. At commercial-scales, geological storage and utilization of CO_2 has widely been applied for Enhanced Oil Recovery (EOR) in North American oil fields in the West Texas Kelly-Snyder and Permian Basin, United States; Weyburn field and Midale field, Canada. Similarly, in recent years CO_2 is injected into tight gas sediments, shale gas, and coal seams for enhancing the recovery [32]. Injection into depleted gas reservoirs have been demonstrated in Gaz de France (K12-B field) in the Dutch side of the North Sea, however benefits of Enhanced Gas Recovery (EGR) requires further demonstration in the future [1].

Saline aquifer, producing and/or depleted oil and gas reservoirs are the two main conventional storage pathways. Injecting CO_2 into oil/gas reservoirs is a more attractive storage option as it has a potential economic incentive due to enhanced hydrocarbon recovery and production. Moreover, site characterizations, injection wells, surface facilities, and information about caprock presence and integrity would be available due to the long history of production. Conversely, injecting CO_2 into saline aquifers would require investment in infrastructure, site characterization, demonstration of caprock sealing capacity, and continuous monitoring throughout the storage project life cycle. Saline aquifers are more accessible due to wide global distribution, and thus estimated to have significantly higher storage capacity [33].

6.3.2 *Oil and Gas Reservoir Exploitation*

Reservoirs are produced under three drive mechanisms: Primary, secondary, and tertiary which is also known as enhanced recovery. Primary recovery mechanism occurs due to depletion of the natural pressure of the reservoir, as oil and gas are withdrawn, and due to pressure support from the underlaying aquifer. Oil recovery from due to natural pressure depletion usually ranges from 10 to 20% of the Original Oil In Place (OOIP). In the case of pressure support from the aquifer system, oil recovery can range 50–60% or sometimes higher. Secondary recovery mechanisms are utilized in the absence of a natural aquifer, and it involves injection of water or gas into the reservoir as means of reservoir pressure support. These secondary drive mechanisms can enhance the primary recovery (natural pressure depletion) to 30–50%. Injection of water and gas maintains reservoir pressure, sustains production levels, and improves the oil mobility towards the producing wells, as the oil is pushed by the injectant gas and/or water to achieve irreducible oil saturation. Conversely, gas reservoirs have higher primary recovery reaching up to 90% of the initial gas in place (GIIP). This is because of the very low viscosity of the gas which eases fluid mobility. Moreover, gas is highly compressible, so slight pressure depletion will result in gas expansion which drives fluid flow [33].

6.3.3 *Enhance Oil and Gas Recovery*

After the primary and the secondary recovery, tertiary recovery mechanisms are utilized to further improve the oil or gas recovery. Most common techniques are stream injection, which is applied to heavy and highly viscous crude oil. Due to uncertainties and low efficiency in some reservoirs, thermal injection can be unconventionally combined with a chemical injection such as surfactants, polymers, and nanoparticles to further improve the recovery [34]. Another commonly practised EOR technique is the CO_2 injection which can be classified as miscible and immiscible CO_2 flooding. Miscibility of the CO_2 with the crude oil depends on fluid properties, reservoir pressure and temperature. At reservoir depths greater than 800 m, hydrostatic pressure approaches the critical pressure of CO_2 (7.38 MPa), and in the presence of crude oils with a density less than 0.9 at reservoir temperature of 15°C, super critical CO_2 dissolution into the crude oil starts to enhance the mobility of the oil [33]. Therefore, immiscible CO_2 improves oil recovery by pushing the oil to production well (Figure 3c). Conversely, miscible CO_2 flooding reduces oil viscosity, density, interfacial tension, increase oil volume because the CO_2 dissolution will cause a swelling effect and mobilizes the lighter components of the crude oil [35].

From CO_2 storage perspective, the injection of CO_2 at reservoir depths greater than 800 m, at pressures of 10–15 MPa, depending on crude oil composition and the reservoir temperature, the supercritical CO_2, will become fully miscible thus enhancing the oil recovery and enabling higher CO_2 storage efficiency. Figure 3(a) illustrates the miscible CO_2 injection process, whereby the produced CO_2 and crude mix is separated at the surface, resulting in oil, gas, and CO_2. After the CO_2 is separated from the production stream, it can be recycled and utilized for re-injection. When geological storage is incorporated into an enhanced oil and gas recovery project, the main sealing mechanism would be structural trapping of the caprock, or any structural seals present in the original hydrocarbon accumulation. Moreover, CO_2 can dissolve in the residual oil, hence enhancing its mobility towards the production well. CO_2 can also be contained due to solubility trapping after long periods of contact with residual formation water, or the underlaying aquifer. However, in the case of EOR, dissolution of CO_2 into the underlaying aquifer might be restricted due to the presence of oil column between the injected CO_2 phase and the aquifer [33].

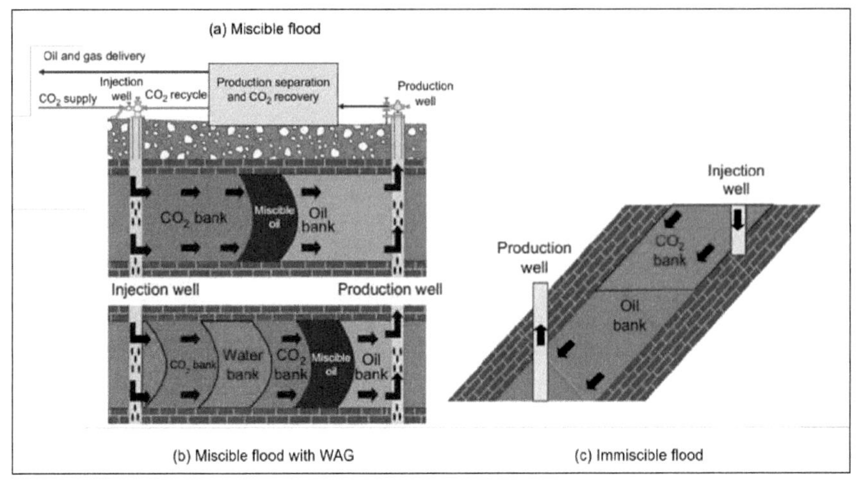

Figure 3. Enhanced Oil Recovery Configurations for CO_2 flooding: (a) Miscible displacement (b) Miscible displacement with water alternating gas; (c) Immiscible flood (Modified from [33]).

Figure 3c illustrates a Water-Alternate-Gas (WAG) injection scheme. There are two types of WAG, miscible and immiscible injection. In miscible WAG, continuous streams of CO_2 with alternating water injection periods are carried out. The objective is to recover more oil compared to pure water flooding [36]. If the injectant CO_2 for the EOR operation is purchased, the project cost can be reduced by decreasing the amount of CO_2 injection and conducting immiscible WAG. Hence, part of the injected CO_2 dissolves with the oil, reducing the oil viscosity and improving the mobility. CO_2 gas volume can further be reduced by separating and recycling the CO_2 at the surface facilities. Eventually, at the end of the economic production of the reservoir, continuous CO_2 can be injected into the depleted reservoir rather than the WAG scheme, depending on project objectives, price, and amount of CO_2 available. Therefore, availability of CO_2 at zero or negative cost in terms of carbon credits, shall incentivize, CO_2 capture, transportation operations, and commencement of large-scale CO_2 utilization for EOR and storage in field that has been considered uneconomical [33].

6.3.4 *Enhanced Gas Recovery (In Depleting and Unconventional Gas Reservoirs)*

Similar to conventional oil reservoirs, gas reservoirs have extensive geological characterization data, and proven ability to contain gases due to the presence of a caprock. Hence, injecting CO_2 into depleting gas reservoir approaching abandonment pressure will provide pressure support, that would allow enhanced gas recovery EGR through the sweep and subsequent production of the residual gas [33]. Hence, injection of supercritical CO_2 into the depleting gas reservoir enhances the sweep of the residual methane due to the density and viscosity difference between these two fluids. Stability of displacement front however, depends on injection and production well configurations. Filling substantial sections of the reservoir volume with CO_2 would be possible before CO_2 starts to breakthrough into the production wells. K12-B was a CO_2 injection project into gas reservoir in Dutch section of the North Sea. This demonstration project showed enhanced gas recover, and storage of 1.0 × 105 tons of separated and captured CO_2 from for more than 13 years the storage [37]. Therefore, this field-scale was the most successful demonstration of EGR via CO_2 injection, and it is a guideline for future EGR projects. Other CO_2 utilization for EGR deals with unconventional resources such as tight gas sands [38], shale gas [39], coal bed methane [40], and lastly methane gas hydrate (which will be discussed in Section 6.5.4).

6.3.5 *Storage in Saline Aquifers*

Injecting CO_2 into depleted oil fields, beneath the Oil-Water Contact (OWC) has been proposed as a pathway for geological storage. Basically, this involves the injection of CO_2 into a saline aquifer, and the injected fluid is structurally trapped as it migrates upward to the original sealing structure (impermeable cap-rock) that contained the hydrocarbon accumulation. Effectiveness of the CO_2 sequestration in saline aquifers depends on the different trapping mechanisms described earlier in (Section 6.1.2). Some of the injected CO_2 will be trapped in the pores and potentially mobilizing some residual oil upwards to the formation crest in the case of a depleted reservoir, however this will take a longer timescale than conventional EOR [33]. Some parts of the injected CO_2 will dissolve in the formation brine, while other parts will react with rock minerals, leading to eventual mineralization of the CO_2. Some important aspects of CO_2 storage projects in saline aquifers include volume of storage site, cap-rock integrity, injection scheme and performance, CO_2 leakage potential and monitoring strategies, environmental impact assessments, and economic incentives for undertaking the CCS project [41, 42].

CO_2 storage in saline aquifer is a mature and considered feasible technically due to the history of ongoing and planned commercial-scale projects globally. It could be economically feasible under the right regulations and economic incentives. Moreover, saline aquifers are widely distributed globally and have high formation rock porosity and permeability which would enable higher storage capacity and ease of injectivity, further necessitating lesser number of injection wells and ease of pressure dissipation within the storage formation [41]. However,

Table 1. Some offshore CO_2 storage and utilization projects [5].

Country	Project	Operation date	Storage type	Reservoir type	Transportation type	Capacity (Mt/year)
Norway	Sleipner	1996	Geological storage	Saline aquifer, depleted reservoir	Ship & Pipeline	Total > 20 Mt
	Snohvit	2008				Total 0.7
	Northern Lights	planned		Saline aquifer		Phase A: 1.5; Phase B:5
UK	Acorn	planned	Geological storage	Depleted O&G field	Ship & Pipeline	0.3
	Net Zero Teesside (NZT) and Zero Carbon Humber (ZCH)	planned		Saline Aquifer	Pipeline	450
	HyNet	planned		Depleted O&G field	Ship & Pipeline	10
	CO2SeaStone, Carbfix	2022	Mineralization	Basalt reservoir	Ship	0.004
Netherlands	K12–B	2004	Geological storage	Gas reservoir	Pipeline	0.1
	Porthos	planned		Depleted gas reservoir	Pipeline	37
Denmark	Greensand	2021 (Pilot phase)	Geological storage	Sandstone reservoir	Ship	by 2025: 1.5 by 2030: 8
China	Enping	2021	Geological storage	Saline aquifer	In-situ separation injection	0.3
Japan	Tomakomai	2016	Geological storage	Saline aquifer	Pipeline	0.3
Australia	CarbonNet	planned	Geological storage	Depleted O&G reservoir	Pipeline	5
	CStore1	planned	Geological storage	Saline aquifer	Ship	1.5–7.5

South Korea	Donghae Gas Field	2022	Geological storage	Depleted gas reservoir	Ship & Pipeline	10
USA	Houston Ship Channel	planned	Geological storage	Depleted O&G reservoir	Pipeline	100
	Timbalier Bay	1984	CO2-EOR	Oil reservoir	Pipeline	Total 0.01
	Quarantine Bay	1981	CO2-EOR	Oil reservoir	Ship	Total 0.028
Brazil	Lula	2011	CO2-EOR	Oil reservoir	FPSO	1.0

salt precipitation is one of the problems faced during CO_2 injection into saline aquifers [43]. Whereby injected dry CO_2 evaporates the water at the injection point, resulting in salt crystallization, reduction in the CO_2 injection rate due to reduced porosity and permeability. Therefore, some methods used to reduce reservoir impairment due to salt precipitation include injection of CO_2 saturated with water vapour to reduce and/or eliminate the evaporation of the formation water. Another cost-effective method is to inject low-salinity water slug before the CO_2 injection, to weaken evaporation and to dissolve any precipitated salt near the injection well-bore.

CO_2 storage into a saline aquifer at capacity of 1 Mt-CO_2/year was demonstrated by Equinor in its continuing projects in Sleipner and Snohvit fields in the North Sea. These CO_2 storage operations are necessary to evade venting CO_2 removed from natural gas production. Currently, several projects are being planned worldwide, and Table 1 provides a summary of some CO_2 storage projects. Comprehensive and up-to-date information about global CCUS projects is maintained by the U.S. Department of Energy (DOE), National Energy Technology Lab (NETL) and Global CCS Institute. Other than EOR, and carbon credits, CO_2 Plume Geothermal (CPG) system has been proposed to generate revenues by coupling CO_2 injection with geothermal energy extraction from deep saline aquifers [33, 44, 45].

6.4 Fundamental Perspective of Hydrate-based CO_2 Storage (HBCS)

Carbon dioxide (CO_2) storage in gas hydrate form under subsurface conditions is a potential approach that proposes an injection of CO_2 into sediments in permafrost locations and/or marine environments, that has thermodynamic condition favourable for converting CO_2 into hydrate. This method could potentially store large quantities of CO_2 in a solid form. Currently, hydrate-based CO_2 storage is at a conceptual stage. Many experimental and simulation studies have been conducted in the past decade that investigate factor that influence hydrate formation, dissociation, and stability. Therefore, this section will explore fundamentals of CO_2 hydrate formation in porous media. Subsequent sections will discuss hydrate-based CO_2 storage pathways, field-scale trials, perspectives on development of these methods, and the future research direction.

6.4.1 *Introduction to Hydrate Basic Concepts*

This section explores the kinetics and thermodynamics formation of CO_2 hydrates in porous media for CO_2 storage application. Table 2 and

Table 2. Summary of reported CO_2 hydrate studies in porous media.

Reference	Study type	Type of porous media	Additives	Results and Limitations
[60]	THS/KHS	Na-MMT	H_2O +NaCl	Clay minerals and salts thermodynamically inhibit hydrates. No study on hydrate dissociation.
[47]	KHS	Silica gel	SDS	SDS enhances hydrate formation uptake and increases the initial hydrate formation rate. No study on hydrate dissociation.
[51]	KHS	Silica gel	Tween-80, DTACl, and SDS	The larger surface area of silica gel resulted in higher gas consumption. SDS was the most effective in enhancing the rate of hydrate formation. No study on hydrate dissociation with brine.
[68]	THS	Silica gel	DI	An increase in pore diameter reduces the water activity in silica gel pores resulting in an inhibition effect. No brine.11
[69]	KHS	Kaoline and Na-MMT	DI	Bentonite thermodynamically inhibits CO_2 hydrates. The hydrate formation in kaolinite is controlled by heat transfer. No brine.
[61]	KHS	Silica sand, Zeolite	SDS	SDS promotes hydrate formation in silica sand than zeolite. No study on hydrate dissociation with brine
[62]	KHS	Silica sand	DI	Suitable CO_2 injection method increases CO_2 hydrate storage capacity in porous media. No brine
[63]	KHS	Quartz sand	DI	CO_2 hydrate Storage capacity increases with decreasing particle size. No brine.
[57]	KHS	Silica sand	Seawater	Gas consumption is higher for smaller size silica particles. Salts act as thermodynamic inhibitors by lowering gas consumption. No study on hydrate dissociation.

Table 2 contd. ...

...Table 2 contd.

Reference	Study type	Type of porous media	Additives	Results and Limitations
[63]	KHS	Quartz sand	DI	Porous media with a pore size of 13.8 nm has maximum gas storage capacity and average formation rate. The smaller the pore size of the porous media, the larger the gas storage capacity. No study on hydrate dissociation with brine.
[59]	KHS	Pumice, FHRC and Silica sand	DI	Pumice showed better hydrate formation kinetics than FHRC. Kinetics was enhanced with a decrease in bed height. Smaller particle size enhanced hydrate formation kinetics. No study on hydrate dissociation with brine.
[54]	THS	Soda glass, mix glass beads and silica gel	DI	Decreasing pore size thermodynamically inhibits hydrates due to the capillary effect. No brine.
[53]	KHS	Cellulose foam	DI	Lower water saturation forms more hydrates than higher saturation. No brine.
[70]	KHS	Glass beads	DI	Hydrates are formed in two stages-hydrate enclathration and continuous occupancy. The gas occupancy stage lasted longer than the previous continuous stage. No study on hydrate dissociation with brine.
[58]	KHS	Toyoura sand	NaCl	The initial rate of hydrate formation is higher in saltwater. Slightly less water conversion to hydrate was found in saltwater compared to pure water. No study on particle size effect on hydrate formation and dissociation kinetics.
[71]	KHS	Silica sand and Silica gel	DI	Water conversion was higher with silica sand compared to silica gel. No study on brine.

[64]	KHS	Glass beads	DI	Flow rate is an important parameter for CO_2 hydrate formation. An optimum flow rate is required for high CO_2 storage. Low operating pressure delays hydrate formation. No study on brine.
[72]	THS/KHS	Glass beads	THF SDS	Increasing THF concentration increases the driving force for hydrate formation with reduced induction time. THF decreases the hydrate phase equilibrium pressure drastically. No study on brine.
[72]	KHS	Glass beads	THF	High operating pressure enhances hydrate formation. No study on brine.
[65]	KHS	Silica gel	DI	The rate of hydrate formation is related to the driving force and the formation rate is fastest in 100nm silica gel pores. No study on hydrate dissociation with brine.
[66]	THS	Silica gel	DI	The pore size of silica gel thermodynamically influences CO_2 hydrate formation. No study in brine.
[52]	KHS	Silica sand, silica gel, Metallic packing	SDS	Metallic packing exhibits high gas uptake and water to hydrate conversion. No study on hydrate dissociation with brine.
[73]	KHS	Frozen quartz sand	DI	The rate of hydrate formation increased with an increase in the initial pressure. No study on hydrate dissociation with brine.
[74]	KHS	Silica sand	C_5H_{10}	FBR showed a higher hydrate formation rate and more gas uptake than UTR. No brine.
[75]	KHS	Na-MMT and UBS	NaCl	Induction time in Na-MMT was shorter than in others. However, hydrate nucleation was faster in Ulleung Basin Sediment (UBS). No study on hydrate dissociation.

Table 2 contd. ...

...Table 2 contd.

Reference	Study type	Type of porous media	Additives	Results and Limitations
[76]	THS	Glass beads	Seawater	Increase in MCP concentration increases the gas uptake and hydrate equilibrium temperature. No study on particle size effect.
[77]	THS/KHS	Silia gel	THF,EGME, SDS and TBAB	THF and SDS promote the gas uptake with increasing driving force. No study on hydrate dissociation with brine.
[78]	THS	Silica gel	DI	Increasing CO_2 concentration shifts the N_2 hydrates equilibrium curve to lower pressures. No brine.
[48]	THS	Glass beads	DI	Gas mixtures with higher CO_2 mole fraction show lower equilibrium pressure under the same temperature. No brine.
[50]	THS	Glass beads	DI	Increasing CO_2 concentration shifts the N_2 hydrates equilibrium curve to lower pressures. No brine.
[79]	THS/KHS	Silica gel	DI	Silica gel enhanced the rate of hydrate formation. No brine.
[72]	THS	Glass beads	SDS	THF decreased the hydrate phase equilibrium pressure. SDS is the best additive showing higher hydrate saturation at 1000 mg/l. No brine.
[80]	THS	Metal-organic framework	DI	Gas hydrates formed only in the meso and macropores and showed slight inhibition compared to the bulk phase. No brine.
[81]	THS	Silica gel	DI	Silica gel enhances the rate of hydrate formation and water to hydrate conversion compared to bulk water. No brine.

[82]	THS	Crystalline swelled Na-MMT	DI	Gas hydrates formed in the clay show both promotion and inhibition effects on the hydrate phase boundary curve. No brine.
[83]	THS	Quartz sand and silica gel	NaCl	The dissociation curve of mixed CO_2+CH_4 hydrate in porous media shifted to the left compared to bulk hydrate. Equilibrium pressure decreased with increasing CO_2 concentration in the gas phase at a constant temperature.
[49]	THS/KHS	Soda glass and silica gel	THF and SDS	THF improves gas consumption at 3 mol%. With increasing driving force, gas consumption increases. No brine.
[84]	THS	Silica glass	DI	Hydrate dissociation temperature was affected by pore diameter. Higher capillary pressure resulted in decreased decomposition temperature. No brine.
[85]	THS	Na-MMT, Kaolinite, and Illite	Glycine, glucose, and urea	Sodium montmorillonite affected CO_2 hydrate phase equilibrium while Kaolinite and Illite showed a negligible effect. Organic matter inhibits the CO_2 hydrate phase equilibrium.

Figure 4 outline hydrate-based CO_2 storage studies in different types of porous media. Moreover, studies often consider thermodynamics and kinetics of different hydrate-forming gases; temperature and pressure conditions; formation and dissociation conditions. Kinetics studies are performed to understand time-dependent behaviour of hydrate formation or dissociation process in porous media. Conversely, thermodynamics studies investigate favourable conditions that lead to CO_2 hydrate formation. The outcomes of these investigations are often used to decide optimal pressure, temperature, additive concentrations that would favour CO_2 storage in porous media and on large-scale would provide valuable insights for storage site and sediment depth selection.

To effectively achieve hydrate-based CO_2 storage, high pressure and low temperature conditions are required in the porous sediments. These conditions can be encountered beneath marine sediments and permafrost locations. Sandstones are practical sediments that can store high CO_2 hydrate quantities due to its high porosities. Sandstones mainly contain minerals such as quartz and felspar which are known as silicates. Other minerals such as clays and shale layers can be found interbedded with sandstone layers in marine sediments. Clays minerals can be cementing agents in the sand sediments, they have small particles sizes, and they are prone to swelling due to their high hydration capacity. Higher concentrations of these minerals in sediment can significantly reduce the local porosity and permeability. Sandstone particles distributions range from 62.5 μm to 2000 μm [46]. Generally, different studies consider porous media minerals present in sandstones [47–50], in order to mimic real sediment compositions, since acquiring actual core samples from hydrate-bearing reservoirs could be challenging due to involved expenses and because of rarity of field-scale hydrate reservoir investigations and operations. Therefore, researchers commonly use porous media such as quartz sand, silica gel, bentonite, kaoline clay, Toyoura sand, glass beads, cellulose foam, and metallic packing in various lab-scale investigations to determine the behaviour of gas hydrates [51–55].

Furthermore, in marine environments such as the Indian Ocean and the South China Sea (SCS), hydrate-bearing sediments are composed of different minerals such as calcite, clays, and silicates in various proportions. Therefore, CO_2 hydrate formation studies in such mixed porous media are currently not explored in the literature. Also, investigations on actual marine sediment outcrops (sediment core samples) could provide realistic hydrate formation potential for any future planned deployment of large-scale hydrate-based CO_2 storage. Currently, research selects porous media to achieve suitable rock properties mainly porosity. Different sediment porosities are acquired from using sediments with various particle size distributions. Earlier studies mainly considered particle

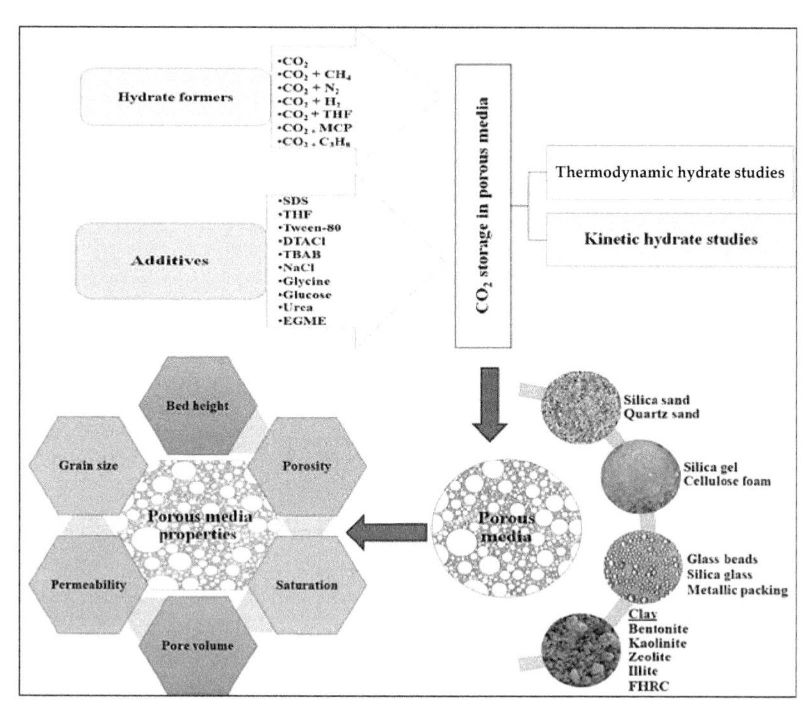

Figure 4. Summary of CO_2 hydrate studies in different porous media systems (reproduced with permission from [56]).

ranges between 0.0007 μm and 1000 μm, which yield sediment porosities ranging from 15 to 56%. Moreover, some studies consider different level of water saturation within the sediment and the state of injected gas.

Typically, the main gas hydrate formers that are usually studied include: pure CO_2 [55, 57–59], [52, 60–64]; the effects of binary mixtures of (CO_2 + CH_4) [65, 66], (CO_2 + N_2), and other guest molecules including: C_3H_8, C_2H_6; TBAB, THF, and cyclopentane [67]. Depending on the present guest molecules or mix-gases, the experimental temperature and pressures ranged (1.13–10.50 MPa) and (269.65–289.95 K), respectively. These experimental studies included other guest molecules in order to investigate their effect on CO_2 hydrate formation and dissociation behaviour in porous media. This occurred in two different ways, first by shifting the CO_2 hydrate equilibrium curve to favourable pressure and temperature conditions; and secondly by enhancing the onset of CO_2 hydrate nucleation and subsequent hydrate growth kinetics. Furthermore, other studies consider the effect of hydrate promoters such as Sodium Dodecyl Sulphate (SDS), different amino acids, Tween-80, urea, glycine, and glucose. These studies have been conducted in saline conditions (with brine) to simulate salinity of seawater in marine sediments. Most additives are studied at concentration lower than 20 wt.%, because the usage of low-dosage additives would be more economical than applying excessive chemical concentrations to achieve better CO_2 hydrate formation results. Typically, hydrate autoclaves with stirrers, rocking cells, batch reactors that can hold sediments are utilized for experiment studies of hydrate kinetics and thermodynamics. Some advanced laboratory setups incorporate Magnetic Resonance Imaging (MRI), Nuclear Magnetic Resonance (NMR), Raman spectroscopy, and Gas Chromatography (GC) for further analysis of hydrate behaviour.

6.4.2 CO_2 Hydrate Phase Behaviour in Porous Media

Estimating CO_2 hydrate phase equilibrium condition in various porous medium is significant for planning and selection of suitable CO_2 storage sites that exhibit favourable hydrate formation condition in terms of sediment pressure and temperature. Moreover, effects of factors such as type of porous media on phase behaviours is important for building a clear understanding regarding hydrate formation. For instance, researchers usually investigate sediment properties such as particles size distribution, mineral composition, salinity, and effect of additives on equilibrium conditions of CO_2 hydrates. All these factors can influence the amount of CO_2 hydrate that could form in the storage sediment. Furthermore, knowledge of hydrate phase behaviour in particular sediment would determine the Hydrate Stability Zone (HSZ), depth of CO_2 injection, and margins of the storage sediment. Conversely, CO_2 phase behaviour is often investigated to evaluate CO_2 separation efficiency in fuel gas systems for carbon capture applications [48, 79]. For CO_2 storage application and in the presence of porous media, determining gas hydrate phase behaviour under different pressure conditions and influences of additives is relevant for CO_2-CH_4 gas hydrate replacement technique, which is mainly driven by chemical potential difference between two gas hydrate systems and their equilibrium conditions. This technique is further discussed in (Section 6.5.4). Different types of porous media have varying particle size distribution which influences hydrate phase boundary [66]. Sediment properties are inter-related, whereby particle sizes govern capillary effect on injected fluids and their mobility, and surface area [86]. Porosity and permeability of the hydrate-bearing sediment could influence hydrate formation behaviour [87]. Therefore, clear understanding of the effects of these factors would enable an accurate prediction of hydrate formation in different conditions and storage mediums particularly marine sediment.

6.4.3 CO_2 Hydrate Formation Kinetics in Porous Media

Hydrate kinetic studies are important to understand time-dependent behaviour of CO_2 hydrate formation in the presence of different porous media. Kinetic indicators such as nucleation time, hydrate formation rate, gas, and water to hydrate conversion ratio can be used to determine the storage efficiency of CO_2 in hydrate form. Nucleation time provides time elapsed before hydrate starts to form in a system at hydrate equilibrium conditions. Rate of hydrate formation can be studied to determine if a particular sediment is favourable for hydrate expansion as it grows in size within the porous media. CO_2 to hydrate conversion ratio considers the amount of CO_2 gas molecules consumed after hydrate nucleation. Therefore, by analyzing these factors, one might be able to establish the effects of external factors on hydrate formation kinetics and also recommend favourable conditions that hasten hydrate formation which will be beneficial in the context of large-scale deployment of hydrate-based CO_2 storage. Disregarding the influence of experimental conditions including pressure, temperature, additive concentrations, porous media type, gas hydrate kinetics can be apparatus dependent. Hence matching results might not be obtained from different apparatus. However, general trends can be obtained after several repetitions of each experiment. Other sources of uncertainties in hydrate kinetic results could be from the reactor type, shape, wall roughness, and gas injection rates.

Conversely, experimental aspects that affect hydrate formation kinetics in porous media include particles size distribution, surface area, water saturation, sediment-bed height, type, and concentration of additives. Therefore, uncertainties that arise from an experimental setup coupled with the stochastic nature of hydrate formation, any experimental errors such as improper utilization of equipment and/or inaccurate data collection approaches, would eventually result in flawed results and subsequently lead to challenges while predicting behaviour of hydrate formation kinetics. Hence, to ensure accurate outcomes from kinetic experiments, each factor must be studied carefully while keeping other factors constant, and multiple repetition experiments are recommended, while avoiding memory effect of each experiment. Therefore, the presence of porous media, is known to affect the phase behaviour of hydrate formation, especially pressure at which hydrates start forming. On the other hand, induction time in porous media can be facilitated by high surface area provided by small particle-size sediments. This is because hydrate starts to nucleate on these surfaces quicker, which is known as heterogenous nucleation. In the absence of stirring/ agitation mechanism in porous media experiments, hydrate growth is typically delayed. Figure 5 illustrates common porous media properties that influence kinetics of hydrate formation and dissociation.

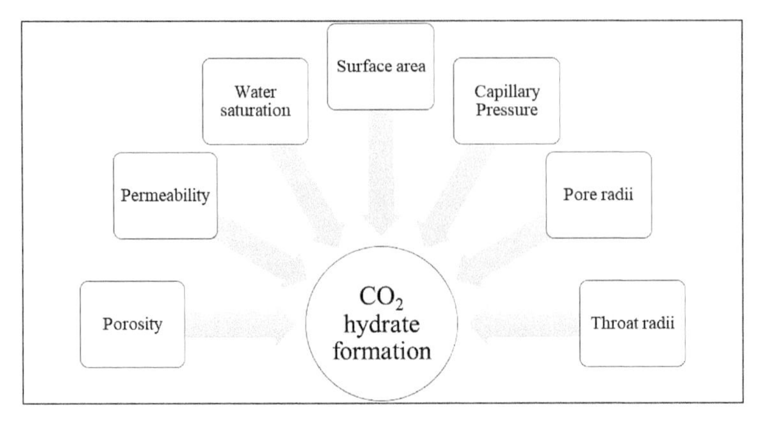

Figure 5. Porous media properties that affect CO_2 hydrate kinetics (reproduced with permission from [56]).

6.4.4 CO_2 Hydrate Dissociation and Stability in Porous Media

Long-term fate of the injected CO_2 and the stability of the CO_2 hydrate in the storage sediment are crucial aspects that require continuous monitoring. Current research investigates the stability of CO_2 hydrate in marine sediment conditions. Due to the consequential environmental impacts of CO_2 leakage into marine environments, further research is always encouraged to establish factors that destabilize or enhance the integrity of CO_2 hydrates within the marine sediments. Moreover, typical storage sites that favour hydrate formation occur in shallow sediments within 300–600 metres beneath the seabed, as that depth of water provides additional hydrostatic pressure that would allow the sediments favour hydrate formation. However, shallow sediments can be prone to perturbation from tectonic activities, drilling operations, or extreme oceanic currents. These factors might affect the *in-situ* sediment pressure or temperature, which could destabilize hydrate-bearing sediments. Currently, field-scale investigations of these factors on hydrate-bearing sediments have not yet been conducted. Therefore, lab-scale researchers study the stability of hydrate-bearing porous media, mimicking real oceanic conditions. Table 3 summarizes some experimental studies that consider hydrate stability in porous media under different conditions.

Methane hydrate-bearing sediment in the permafrost and Arctic environments have been releasing methane gas because of global warming. Hence, dissociation of methane hydrates can be used to understand possible dissociation behaviour of CO_2 hydrates in storage sediments [91–93]. According to Yang et al. [58], saline conditions reduces time taken to release an arbitrary amount of CO_2 by half compared to CO_2 released from hydrate dissociation in porous media saturated with pure water. The effects of different sediment compositions such as presence of

Table 3. Summary of CO_2 hydrate dissociation in different porous media systems.

Reference	Type of study	Type porous media	Type of gas	Remarks
[71]	Kinetic hydrate study	Silica sand and silica gel	CO_2/H_2	The study was conducted in deionized water and not brine solution.
[88]	Kinetic hydrate study	Glass beads	CO_2	The study was conducted in deionized water and not brine solution.
[89]	Kinetic hydrate study	Glass beads	CO_2	The study was conducted in deionized water and not brine solution.
[58]	Kinetic hydrate study	Toyoura sand	CO_2	The effect of varying particle size on the dissociation kinetics of CO_2 hydrates is not studied in the brine solution.
[64]	Kinetic hydrate study	Glass beads	CO_2	The study was conducted in deionized water and not brine solution.
[90]	Kinetic hydrate study	Consolidated and unconsolidated sandstone	CO_2	The study was conducted in deionized water and not brine solution.

clay minerals have not been investigated in the literature. Furthermore, petrophysical properties of sediments such as particle size distribution, porosity, permeability affects the degree of hydrate dissociation. These physical properties greatly influence the heat transfer within the hydrate-bearing sediments, which subsequently affects the hydrate dissociation rate [92]. Excessive heating, and porous media with high thermal conductivity would result in abrupt hydrate dissociation [88]. Moreover, sediments with particle side distribution of (35–60 mesh size) exhibit weak stability with high hydrate dissociation rate, in contrast with particle sizes smaller than 35 and larger than 60 mesh [90]. Hence, porous media with higher porosities could provide a stable hydrate storage, due to the low thermal conductivity of these sediments which reduces hydrate dissociation rate [94]. Moreover, hydrate dissociation in porous sediments is influenced by the *in-situ* hydrate surface area [95]. This means that highly saturated hydrate-bearing deposits could be susceptible to continuous dissociation, in contrast with scattered hydrate deposits. Dissociation kinetics of hydrate in sediments can also be influenced by the presence of additives such as brine, amino acids, and surfactants. Therefore, understanding sediment properties that influence thermal conductivity, dissociation kinetics, and hydrate stability would enable proper evaluation, screening, and selection of suitable CO_2 hydrates storage sites accounting for hydrate dissociation tendency.

6.4.5 *Application of Hydrate Promoters in Porous Media*

Kinetics of hydrate formation is often slow in natural systems; therefore, chemical additives are used to enhance hydrate formation. Typically, different hydrate promoting chemicals are studied to determine their effect on hydrate induction and growth, mechanism of promotion, and their optimum concentration. Based on their application, the two classes of chemical additives are thermodynamic and kinetic promoters. Thermodynamics promoters enlarge the phase boundary in porous media, which enable hydrate formation in a larger range of pressures and temperatures. In other words, these promoters can reduce pressure and increase at which hydrate starts to form. In terms of CO_2 storage application, this means that CO_2 can be injected at shallower depths, hence extending the storage sediment boundaries to store more CO_2 hydrate. Conversely, kinetic promoters quicken hydrate formation, resulting in shorter nucleation time, higher rate of gas uptake, and hydrate growth. In CO_2 storage application, hastened CO_2 hydrate formation would enable faster formation of the hydrate cap that can seal the CO_2 bulk phase from leaking to the ocean floor (see Figure 8b, in Section 6.5.3). Therefore, experimental investigations that mimic CO_2 hydrate storage in marine sediments use saline conditions using NaCl and different forms of porous media, but mainly silica sand. However, salinity of oceanic water is well known to inhibit hydrate formation and growth, due to the strong electrostatic forces of NaCl ions that interact with water molecules disrupting the formation of water clathrates that encapsulate the gas (hydrate former). Mekala et al. [57] and Yang et al. [58] reported in their studies that in the presence of porous media, salinity due to NaCl promotes hydrate nucleation and formation.

6.4.6 *Kinetic Hydrate Promoters (Low Dosage and Environment Friendly)*

Industrial applications encourage the use of Low-Dosage Hydrate Promoters (LDHP), as these chemical additives are effective at lower concentrations (< 10,000 ppm), which also saves costs. The role of kinetic promoters is to lessen the interfacial tension (IFT) between the CO_2 gas and water phase within the porous media, in CO_2 storage application. Reduction of IFT would further enhance CO_2 solubility in water, which would increase CO_2 to hydrate conversion. Commonly, kinetic promoters are surfactants, amino acids, lignosulphonate, and nano-fluids. Sodium dodecyl sulphate (SDS) is an anionic surfactant that is widely used as a hydrate kinetic promoter. However, due to the toxicity of SDS, its application for promoting CO_2 hydrate storage in marine sediments could have environmental safety concerns towards the marine ecosystems that surround the storage location. Conversely, conventions hydrate promoters have been investigated on their effects on nucleation time with porous media by several researchers [72, 77, 96–99] SDS reduces hydrate induction time by three-folds at 0.4 wt.% SDS concentration. Moreover, in the presence of porous media, SDS also reduces hydrate nucleation time [51]. Despite SDS' good performance with hydrate promotion, SDS generates large amount of foam when hydrates are dissociated, hence in large industrial application, SDS foaming can hamper the process. For instance, injecting SDS as promoter into a methane hydrate reservoir for CO_2-CH_4 replacement might at, however liberated methane gas from the dissociation could be restricted by significant foaming for chemicals such as SDS. Therefore, investigation into bio-promoters that are cost-effective with less or no foaming generation, would contribute to safer deployment of chemical-assisted hydrate-based CO_2 storage.

6.5 Oceanic and Hydrate-based CO_2 Storage Pathways

Oceans cover over 70% of the Earth's surface, providing significantly large theoretical CO_2 storage capacity that ranges from 4000–10,000 Gt-CO_2. That is almost 13 times or approximately 250 years of storage capacity at our current CO_2 global emission rate. Methods that have been studied to increase oceanic CO_2 uptake include application of biological fertilization, chemical approach which involves reducing acidity of oceans to hasten limestone weathering, and physical methods such as CO_2 dissolution, liquid CO_2 injection which forms dense and sinking pools in the deep ocean [100]. Conversely, injected CO_2 in the deep ocean forms CO_2 hydrate due to the high pressure and low temperature conditions in the deep oceans. These hydrates are negatively buoyant and eventually sink to the ocean floor. Therefore, this section explores hydrate-based CO_2 storage pathways that can be classified into four principal approaches: CO_2 injection into permafrost regions; deep oceans; shallow marine sediments; methane hydrate reservoirs for CO_2-CH_4 hydrate replacement.

6.5.1 *CO_2 Hydrate Storage in the Terrestrial Permafrost*

NGHs are commonly found in the permafrost regions. Hydrate resources in the permafrost are estimated to be less than that in marine regions by two orders of magnitude or more. However, NGH recovery from permafrost regions is generally considered simpler and more cost-effective compared to marine environments. This is primarily due to the lower complexity of technical challenges involved in onshore operation [101]. Equally,

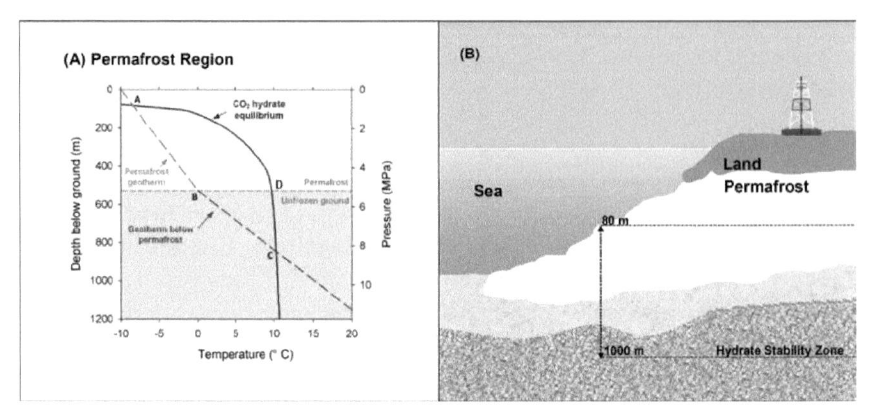

Figure 6. Illustration of Hydrate Stability Zone (HSZ) in a permafrost region; Modified from [103].

injecting and storing CO_2 as gas hydrate beneath the permafrost sediments could be more feasible. Figure 6(B) illustrates hypothetical CO_2 hydrate stability zone (HSZ) with a geothermal gradients for the permafrost layer (1.9°C/100 m) and the unfrozen sediment (3.2°C/100 m) [102]. CO_2 hydrates are stable in depths between points A and C.

Melnikov and Drozdov [104] suggested that favourable CO_2 storage locations should have a permafrost sealing cap, which is a continuous sediment layer that lacks warm/unfrozen ground. However, warm spots and faults are common in permafrost layers. Therefore, the risk of CO_2 leakage must be thoroughly assessed. The feasibility of storing CO_2 beneath the Siberian permafrost has been studied [105, 106]. The HSZ thickness varied from 50 to 800 m at different locations, due to the varying geothermal profiles and permafrost thickness. Because of the low permeability of the frozen sediments in the permafrost layer, it is recommended to sequester the CO_2 in reservoirs beneath the permafrost (region BDC, Figure 6). However, injecting CO_2 into the HSZ below the permafrost will cause eventual porosity plugging due to hydrate formation with unfrozen water, which will impede further CO_2 injection. Therefore, Duchkov et al. [106] suggested CO_2 injection beneath the HSZ at depths greater 600 to 1000 m. This approach would involve the utilization of multiple barriers, including a geological cap composed of mud or shale, an impermeable layer of hydrate-cemented sediment, and the presence of permafrost.

6.5.2 Oceanic CO_2 Hydrate Storage

Figure 7(A) illustrates HSZ for CO_2 hydrates in deep oceans, which is determined by the depth, and superimposed hydrothermal and geothermal gradient [107]. Hence, due to the equilibrium conditions

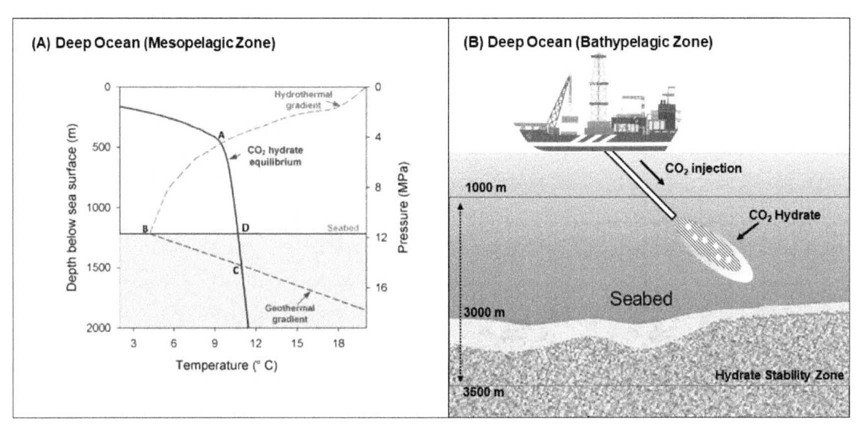

Figure 7. Illustration of Hydrate Stability Zone (HSZ) in deep ocean; Modified from [103].

in region BDA in Figure 7), CO_2 hydrates can form and accumulate. However, these hydrate formations would be exposed to oceanic currents and perturbations which may destabilize the hydrate structures and may cause a probable CO_2 leakage.

In 1999, Brewer et al. [108] conducted the first series of oceanic CO_2 injection field experiments. They injected small amounts of liquid CO_2 (a few litres) into the water column at different water depths (ranging from 349 m to 3627 m) within the hydrate stability zone in Monterey Bay, California. CO_2 hydrate formation was observed at various depths in the test beaker, and thin hydrate films were observed surrounding the liquid CO_2 phase. At the greatest injection depth of 3627 m, the gradual formation of hydrates resulted in the expansion of the CO_2 liquid volume in the beaker, leading to the spillage of CO_2 onto the seafloor. The spilled CO_2 droplets exhibited high mobility, potentially due to dissolution and ocean currents. Therefore, outcomes of this field study suggest that permanent CO_2 hydrate disposal in deep oceans is not feasible due to concerns with hydrate stability and environmental implications of the exothermic nature of hydrate formation and the CO_2 dissolution.

Lee et al. [109] suggested utilizing a co-flow reactor to create a sinking composite consisting of CO_2 hydrate, liquid CO_2, and water. Their co-flow reactor configuration addressed the clogging issue encountered during liquid CO_2 injection by injecting water into liquid CO_2, resulting in the formation of a paste-like composite. This composite could be modified by adjusting the injection pressure and flow rates of water and liquid CO_2. Moreover, a simulation study indicated that the formation of CO_2 hydrates plays a crucial role in achieving the sinking composite particles. However, it should be noted that experimental results obtained using freshwater may differ significantly from those obtained using seawater [109]. Furthermore, in a collaborative effort with Tsouris et al. [110], the same research group conducted a field experiment utilizing a co-flow reactor and successfully formed composite particles consisting of CO_2 hydrate, CO_2 liquid, and seawater at sea depths ranging from 1100 to 1300 m. However, since the composite particles exhibited either neutral buoyancy or slight positive buoyancy, it was necessary to enhance the density of the CO_2 hydrate composite and achieve a negative buoyancy. In the subsequent phase of the field experiment, higher CO_2 to hydrate phase conversion of (~ 40%) was achieved by improving the injector geometry in the co-flow reactor. This resulted in highly dense CO_2 composite particles with (~ 5 cm/s) sinking rate [111]. The following studies scaled-up the hydrate formation reactors [112–114] and developed high-pressure water tunnel facilities to mimic oceanic environments in a laboratory setting [115].

The International Energy Agency (IEA) proposed the production of CO_2 hydrates at onshore location and then transported through shipping to deep oceans, where it will be discharged and sunken to the ocean floor [116]. At offshore production sites, compressed CO_2 hydrate blocks can be made, which would be denser than seawater to allow sinking. Hence, both approaches, involving CO_2 hydrates and composite particles, show promise in terms of their sinking behaviour. Rehder et al. [117] conducted a study to investigate the dissolution rate of compact CO_2 hydrate and CH_4 hydrate cylindrical samples (3 cm by length and 2.2 cm in diameter) on the ocean floor at depth of 1,028 m. They observed that the sample diameter reduced, and dissolution rates were (4.15 ± 0.5 mmol CO_2/m^2s) and (0.37 ± 0.03 mmol CH_4/m^2s). The dissociation kinetics were controlled by diffusion due to oceanic disturbances. They concluded that methane hydrate is more stable and less susceptible to diffusion and dissolution at the investigated depth. Hence, CO_2 hydrate stability in marine environments and its effect on the pristine eco-systems are major concerns that limit CO_2 hydrate disposal in deep sea [118, 119].

6.5.3 CO_2 Hydrate Storage in Marine Sediments

Figure 8(A) illustrates HSZ for CO_2 hydrates in deep oceans, which is determined by the depth, superimposed hydrothermal and geothermal gradient [107]. Hence, due to the equilibrium conditions in region BDC in Figure 8, CO_2 hydrates can form and accumulate. However, these hydrate formations would be exposed to oceanic currents and perturbations which may destabilize the hydrate structures and may cause a probable CO_2 leakage.

Figure 8 illustrates the physical model of CO_2 injection below the hydrate stability zone (HSZ), whereby injected CO_2 migrates upward for the hydrate layer at the bottom and within the HSZ. The hydrate cap prevents continuous upward migration of the CO_2 and leakage.

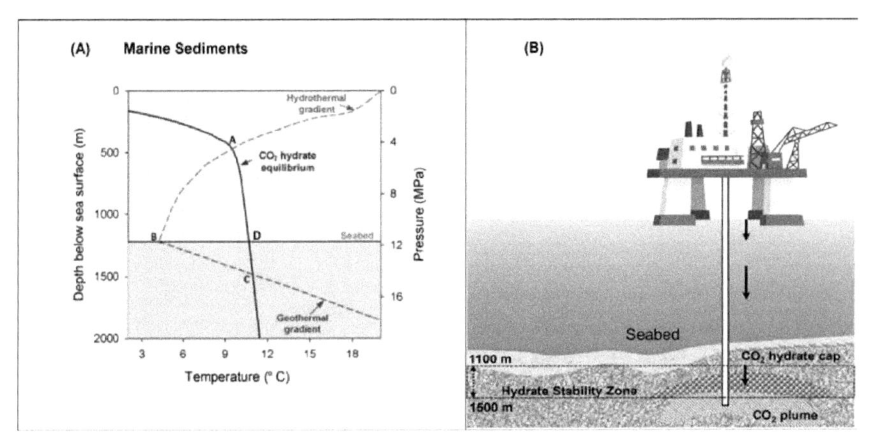

Figure 8. Illustration of Hydrate Stability Zone (HSZ) beneath marine sediments; Modified from [103].

Below this hydrate cap, the gaseous/liquid CO_2 is trapped, and due to gravity, some of the CO_2 will flow downward to the deeper sections of the reservoir. Moreover, CO_2 may also exist as a super-critical fluid beneath the hydrate cap and form a plume that spreads laterally following the *in-situ* pore network of storage sediment. Therefore, high permeability would ease further the gas injection below the HSZ, this is because direct injection of CO_2 into the HSZ would eventually cause hydrate formation near the injection well (injection point), which will ultimately reduce the permeability of the near formation, reduce injection rates, and hinder CO_2 flow into the sediment.

6.5.1 CO_2-CH_4 *Hydrate Replacement*

CO_2 gas is injected into a methane hydrate reservoir to achieve simultaneous CO_2 storage and CH_4 replacement [120]. Hence, this gas exchange phenomenon produces methane and provides significant economic prospects. The technique of gas replacement within the clathrate hydrate structure was initially proposed by Ebinuma [121] and Ohgaki [122] in the mid-1990s. Compared to the previous three production methods (depressurization, thermal stimulation, and inhibitor injection), gas replacement does not involve hydrate phase transition (from solid to water and gas). However, it involves a direct replacement of CH_4 by the injected CO_2 molecules within the clathrate hydrate structure (see Figure 9). Hence, this method provides a stable methane production and CO_2 storage potential [123]. It also prevents dramatic variation in the formation stress which could have caused sediment instabilities, collapse, gas leakage, and other environmental hazards [124]. Figure 10 illustrates that CH_4 hydrate has a higher equilibrium pressure compared to CO_2

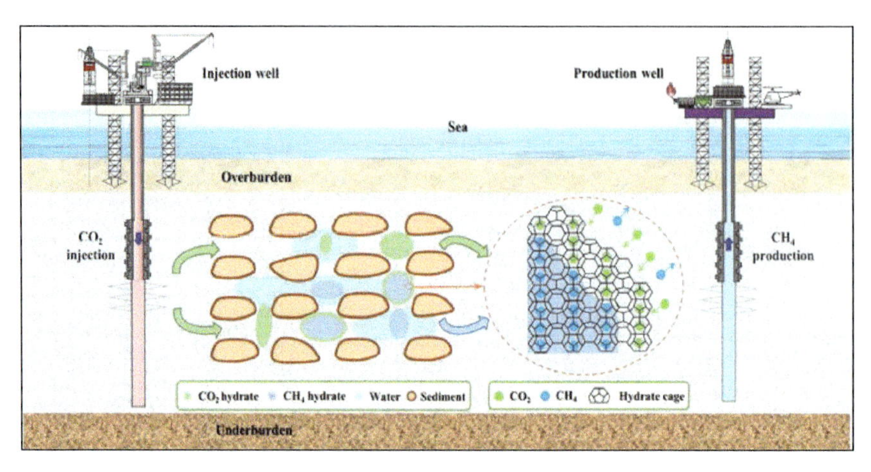

Figure 9. Schematic illustration of CH_4–CO_2 hydrate replacement mechanism (Reproduced with permission from [127]).

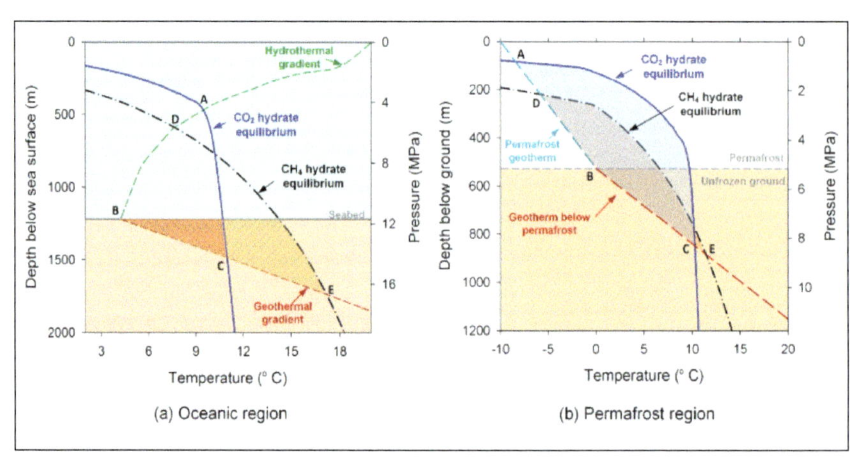

Figure 10. CO_2 & CH_4 hydrate equilibrium curves in oceanic and permafrost regions (Modified from [103]).

hydrate at a constant temperature [120]. Hence, due to this thermodynamic property, CH_4 molecules are replaced by CO_2 molecules. Moreover, the difference in chemical potential [122] and fugacity between CH_4 hydrate and the CO_2 fluid phase provide the driving force for the replacement reaction [125, 126].

Experimental investigations of CH_4-CO_2 replacement showed that the replacement occurred in two stages; rapid surface reaction and a gradual slow reaction due to the formation of high resistance layer of the mixed hydrates, which impedes further CO_2 diffusion and CH_4 replacement [128, 129]. Hence, reaction kinetics and mass transfer are the major limitations of this procedure. Therefore, further studies on CH_4-CO_2 exchange and exploitation of NGHs, will have future economic and environmental significance. Currently, lab-scale research on CH_4-CO_2 exchange mainly includes replacement mechanism, influencing factors, and feasibility studies. Feasibility of CH_4-CO_2 exchange in small-scale marine experiments were established [130]. Moreover, the first large-scale CH_4-CO_2 exchange field test was conducted successfully in the permafrost's of Alaska north slope (ANS) in 2012 [131]. Therefore, factors that impact the replacement efficiency are CO_2 phase, supplementary gas mixtures (CO_2 + N_2 or H_2), thermodynamic promoters, kinetic promoters, injection pressure, depressurization, thermal stimulation, porous media, pore saturation, and hydrate structure.

6.5.4.1 CO_2-CH_4 Hydrate Experimental Studies

Therefore, it is important to experimentally study and comprehend factors that influence the CO_2-CH_4 hydrate replacement efficiency, in order to develop suitable strategies to extract natural gas hydrates NGH via a CO_2 injection. Scientific breakthroughs in lab-scale experimentation would encourage economic incentives for feasibility studies that would ultimately lead to more field-scale deployment of the hydrate replacement technique for CO_2 sequestration and methane recovery. Experimental studies currently explore the effect of CO_2 phase (gas, liquid, or emulsion), supplementary gas, pressure and temperature, effect of porous media, hydrate saturation, and effect of kinetic and thermodynamic promoters. The replacement efficiency is also influenced by other conditions such as: gas initial concentration, injection rate, salt concentration, surfactants, and hydrate replacement strategy.

6.5.4.2 CO_2-CH_4 Hydrate Replacement at Field Scale

In 2012, Conoco Phillips, JOGMEC, and the U.S. DOE conducted a field production trial of Natural Gas Hydrates (NGH) using the CH_4-CO_2 hydrate replacement technique in Prudhoe Bay on the Alaska North Slope (ANS) [131, 132]. The main objective of the field test was to validate the hydrate replacement mechanism identified in laboratory studies [133, 134]. Moreover, the field test aimed to demonstrate CO_2 gas injection into the NGH reservoir, and methane production through a combination of depressurization and exchange while minimizing water and sand production. The target interval was perforated between (683 & 693 m) and was screened for sand control [131, 135]. During the field trial, approximately 6,000 m^3 of a gas mixture consisting of 23% CO_2, 77% N_2, and small amounts of tracers were injected into the hydrate-bearing interval over a two-week period. As the injection progressed, the permeability of the reservoir decreased, leading to a reduction in the gas injection rate.

The permeability variations are mainly caused by changes in hydrate saturation near the wellbore. Analysis using a Hall plot suggested that the formation and dissociation of hydrates could influence the gas injectivity. Following the gas injection phase, the well was shut for four days and then reopened for gas production [131, 135].

The Ignik Sikumi field was produced in two stages, first was unassisted flow and the second was jet pump assisted flow. Moreover, the assisted flow (gas lift via jet pump) phase was carried out in three stages with varying bottomhole pressures (BHPs). These stages maintained bottomhole pressures (BHPs) above, at, and below the equilibrium conditions of CH_4 hydrates. In order to examine the produced gas composition, in-line gas chromatograph was employed. During the unassisted flow phase, gas production was carried out without the use of an artificial lift, and the gas was produced above the CH_4 hydrate equilibrium condition for a duration of 1.5 days. Cumulative gas production was approximately 1,100 m^3 with minimal CH_4 volume fraction. During the first stage of the assisted flow phase, jet pump was utilized, and the gas production was carried out above the equilibrium conditions of CH_4 hydrate. Over a period of 8 days, the well flowed and produced approximately 9,000 m^3 of gas, with a methane composition of around 70 mol%. Operation disruption from sand production was reported twice during this stage. In the second stage of the assisted flow, the jet pump decreased the bottomhole pressure near to methane hydrate equilibrium pressure. This allowed for a higher flow rate, resulting in the production of approximately 10,000 m^3 of gas with a methane composition of around 80 mol% over a period of 2.5 days. At this stage, production ceased due to hydrate reformation within the producing lines. The jet pump was replaced to address the issue of ice/hydrate plugging. Additionally, a small amount of heated glycol was injected below the jet pump to inhibit hydrate formation. At the final stage of the assisted flow, the BHP was lowered below the CH_4 hydrate equilibrium pressure. Approximately 14,600 m^3 of gas production was achieved with a methane composition of around 90 mol% over a period of 19 days [136].

The gas production rate increased progressively and was followed by an increase in water and sand production. Furthermore, the volume fraction of methane gas in the produced gas mixture also showed a gradual increase. The cumulative production volume methane gas was approximately 24,500 m^3 over the 30-day testing period. Moreover, 70% of the injected N_2 gas was successfully recovered, conversely only 40% of the CO_2 was recovered, revealing the reservoir preferential CO_2 retention and the occurrence of bulk CH_4-CO_2 hydrate exchange [131, 132]. This field test provided insights on production behaviour of a NGH reservoir during the CH_4-CO_2 replacement technique. Nevertheless, observed inconsistencies between field production behaviour/data and the prediction simulations could be due to the kinetic dominant exchange process, reservoir heterogeneities, or inadequate heat transfer model [131, 135]. The Ignik Sikumi field production was the first to utilize the gas exchange mechanism for CO_2 sequestration and methane exploitation from a NGH deposits, highlighting the significance flow assurance application and sand control measures to ensure stable, effective, and long-term sustainable methane exploitation from hydrate reservoirs, irrespective of the production mechanism.

6.6 Conclusion and Perspectives

This chapter has discussed conventional and hydrate-based CO_2 storage pathways. Injecting CO_2 in saline aquifers and hydrocarbon-bearing formation to enhance oil and gas recovery are common practices for CO_2 storage and utilization. Currently, hydrate-based CO_2 pathways are of great interest because their development and maturation would lead to significant increments in local and global CO_2 sequestration capabilities in terms of increasing storage capacity available for meeting emission reduction goals for carbon neutral future. Hydrate-based CO_2 storage methods involve injection of CO_2 into permafrost environments, deep ocean, beneath marine sediments, and lastly methane hydrate-bearing sediment. Most studies at present are focused on CO_2 storage into methane hydrate reservoirs because of the economic incentives from procuring methane gas. The main objectives of CO_2 injection into methane hydrate are the simultaneous CO_2 storage and CH_4 recovery. Despite its prospects, CO_2-CH_4 hydrate replacement is facing efficiency challenges, when mixed gas hydrate forms a mass barrier that inhibits further injection of CO_2 gas or production of CH_4 gas. Therefore, further research is encouraged to explore chemical and mechanical approaches to enhance the gas replacement process. Till date, only one CO_2-CH_4 hydrate replacement field-scale demonstration project has been carried out in Iġnik Sikumi gas hydrate field in Alaska North Slope, United States. The demonstration project concluded that injecting mixed gas (N_2 and CO_2) resulted in 70% N_2 and 40% CO_2 recovery, revealing the reservoir preferential CO_2 retention. However, further large-scale demonstration projects are required to assess process efficiencies and economic viability.

Furthermore, injected CO_2 into CH_4 hydrate sediments below the hydrate stability zone can be trapped in different mechanisms such as stratigraphic due to caprock/hydrate cap, and could dissolve in formation brine and eventually mineralize into the rock minerals. Hence further research is required to quantify the distribution of CO_2 entrapment under different mechanisms through experimentation which would lead to physically accurate modelling of CO_2 hydrate storage capacity. More field-scale projects would yield valuable information about the behaviour of the methane hydrate reservoirs in terms of sediment stability, preferential gas retention, gas replacement efficiency under varying conditions, and effects of hydrate additives. Such information would be utilized for enhancing gas injection schemes and proper planning for field development strategies. This would enable efficient exploitation of CH_4 gas from hydrate deposited, which are the most abundant on Earth however lacked technological advancement in recovery techniques.

Currently, experimental research efforts are focusing on implications of different factors on the gas replacement process. Such factors include CO_2 phase, concentration of secondary gas in the injection feed, injection pressure, thermal and chemical stimulation, initial hydrate saturation, effects of sediment properties, and injection of nano-fluids. Particularly, to avoid rapid CO_2 hydrate formation in the vicinity of injection wellbore, it is recommended to avoid excessive injection pressures. Adding secondary gas such as H_2 or N_2 into the CO_2 injection feed would enhance the CH_4 recovery but could reduce the CO_2 storage efficiency if the ratio of the secondary gas is high. Moreover, combining CO_2 or CO_2-mix gas injection with traditional methods such as depressurization, and thermal and chemical stimulation would enhance the gas replacement process. However, a proper strategy of combining these methods should be determined through large-scale demonstration projects, that are conducted for longer periods than that of the Iġnik Sikumi field test. This is necessary to determine long-term gas production and storage data, sand and water production tendencies, and reservoir stability during the operation.

Ultimately, there are several aspects that require further research to enhance the hydrate replacement process and encourage deployment of field-scale tests that would provide confidence for future investment to undertake CO_2-CH_4 hydrate replacement as commercially viable pathway to reduce global anthropologic CO_2 emissions. These aspects are:

- Macro and micro-level studies on hydrate sediment stability during gas injection and production are important to justify confidence in long-term stability of the hydrate reservoir. At the micro-level, hydrate in porous media can be investigated in non-dissociation and dissociation conditions, to determine the factors that would severely impact the gas hydrate stability. Conversely, at macro-level, hydrate morphology as cementing agent can be assessed through formation geo-mechanical stresses. It is important that formation stresses should be stabilized after CH_4 hydrate dissociation and subsequent CO_2 hydrate formation to retain the original sediment stability.
- Sand production and water production events during the CO_2-CH_4 hydrate exchange are common, therefore proper techniques shall be implemented for controlling these events. Moreover, gas replacement

mechanism shall be investigated whether the local CH_4 hydrates dissociate completely, resulting in water and gas production, partial dissociation occurs, or instantaneous gas replacement without hydrate clathrate dissociate. Therefore, developing strategies leading to partial or non-dissociation of local CH_4 hydrates would reduce water and sand production events and ensure sediment stability during the operation.

- Investigation of optimum combination of methane hydrate extraction methods would enhance the gas replacement efficiency. This is because single method approach is less effective. Moreover, deployment of long-term reservoir monitoring would be important for assessing the dynamics of injected CO_2, produced CH_4, leakage potential, and formation of hydrate cementation into a cap-rock.
- Incorporating molecular dynamic simulations with the availability of long-term production and storage data from the hydrate reservoirs would enable accurate simulation of gas replacement mechanisms and *in-situ* effect of promoting and inhibiting hydrate additive, and determine effectiveness of comminating hydrate extraction methods such as CO_2 injection and depressurization.

References

[1] Rackley, S. A. 2017. 11 - Introduction to geological storage. pp. 285–304. *In*: Carbon Capture and Storage (Second Edition), S. A. Rackley Ed. Boston: Butterworth-Heinemann.

[2] Zhang, D. and Song, J. 2014. Mechanisms for geological carbon sequestration. Procedia IUTAM 10: 319–327, 2014/01/01/ 2014, doi: https://doi.org/10.1016/j.piutam.2014.01.027.

[3] Yang, F., Bai, B., Tang, D., Shari, D.-N. and David, W. 2010. Characteristics of CO_2 sequestration in saline aquifers. Petroleum Science 7(1): 83–92, 2010/03/01 2010, doi: 10.1007/s12182-010-0010-3.

[4] Prasad, S. K., Sangwai, J. S. and Byun, H.-S. 2023. A review of the supercritical CO_2 fluid applications for improved oil and gas production and associated carbon storage. Journal of CO_2 Utilization 72: 102479, 2023/06/01/ 2023, doi: https://doi.org/10.1016/j.jcou.2023.102479.

[5] Luo, J., Xie, Y., Hou, M. Z., Xiong, Y., Wu, X., Lüddeke, C. T. et al. 2023. Advances in subsea carbon dioxide utilization and storage. Energy Reviews 2(1): 100016, 2023/03/01/ 2023, doi: https://doi.org/10.1016/j.enrev.2023.100016.

[6] Snæbjörnsdóttir, S. Ó., Sigfússon, B., Marieni, C., Goldberg, D., Gislason, S. R. and Oelkers, E. H. 2020. Carbon dioxide storage through mineral carbonation. Nature Reviews Earth & Environment 1(2): 90–102, 2020/02/01 2020, doi: 10.1038/s43017-019-0011-8.

[7] Sigfusson, B., Gislason, S. R., Matter, J. M., Stute, M., Gunnlaugsson, E., Gunnarsson, I. et al. 2015. Solving the carbon-dioxide buoyancy challenge: The design and field testing of a dissolved CO_2 injection system. International Journal of Greenhouse Gas Control 37: 213–219, 2015/06/01/ 2015, doi: https://doi.org/10.1016/j.ijggc.2015.02.022.

[8] Matter, J. M., Stute, M., Snæbjörnsdottir, S. Ó., Oelkers, E. H., Gislason, S. R., Aradottir, E. S. et al. 2016. Rapid carbon mineralization for permanent disposal

of anthropogenic carbon dioxide emissions. Science 352(6291): 1312–1314. doi: doi:10.1126/science.aad8132.

[9] Pogge von Strandmann, P. A. E., Burton, K. W., Snæbjörnsdóttir, S. O., Sigfússon, B., Aradóttir, E. S., Gunnarsson, I. et al. 2019. Rapid CO_2 mineralisation into calcite at the CarbFix storage site quantified using calcium isotopes. Nature Communications 10(1): 1983, 2019/04/30 2019, doi: 10.1038/s41467-019-10003-8.

[10] Montes-Hernandez, G., Bah, M. and Renard, F. 2020. Mechanism of formation of engineered magnesite: A useful mineral to mitigate CO_2 industrial emissions. Journal of CO_2 Utilization 35: 272–276, 2020/01/01/ 2020, doi: https://doi.org/10.1016/j.jcou.2019.10.006.

[11] Oelkers, E. H., Gislason, S. R. and Matter, J. 2008. Mineral carbonation of CO_2. Elements 4(5): 333–337. doi: 10.2113/gselements.4.5.333.

[12] Clark, D. E., Oelkers, E. H., Gunnarsson, I., Sigfússon, B., Snæbjörnsdóttir, S. Ó., Aradóttir, E. S. et al. 2020. CarbFix2: CO_2 and H2S mineralization during 3.5 years of continuous injection into basaltic rocks at more than 250°C. Geochimica et Cosmochimica Acta 279: 45–66, 2020/06/15/ 2020, doi: https://doi.org/10.1016/j.gca.2020.03.039.

[13] Neeraj and Yadav, S. 2020. Carbon storage by mineral carbonation and industrial applications of CO_2. Materials Science for Energy Technologies 3: 494–500, 2020/01/01/ 2020, doi: https://doi.org/10.1016/j.mset.2020.03.005.

[14] Koide, H., Takahashi, M., Tsukamoto, H. and Shindo, Y. 1995. Self-trapping mechanisms of carbon dioxide in the aquifer disposal. Energy Conversion and Management 36(6): 505–508, 1995/06/01/ 1995, doi: https://doi.org/10.1016/0196-8904(95)00054-H.

[15] Circone, S., Stern, L. A., Kirby, S. H., Durham, W. B., Chakoumakos, B. C., Rawn, C. J. et al. 2003. CO_2 hydrate: synthesis, composition, structure, dissociation behavior, and a comparison to structure I CH_4 hydrate. The Journal of Physical Chemistry B 107(23): 5529–5539, 2003/06/01 2003, doi: 10.1021/jp027391j.

[16] Rackley, S. A. 2017. 13 - Fluid properties and rock–fluid interactions. pp. 337–364. *In*: Carbon Capture and Storage (Second Edition), S. A. Rackley Ed. Boston: Butterworth-Heinemann.

[17] Blunt, M. J. 2017. Multiphase Flow in Permeable Media: A Pore-Scale Perspective. Cambridge: Cambridge University Press.

[18] Lake, L. W. 1989. Enhanced Oil Recovery.

[19] Murphy, Z. W., DiCarlo, D. A., Flemings, P. B. and Daigle, H. 2020. Hydrate is a nonwetting phase in porous media. Geophysical Research Letters 47(16): e2020GL089289, doi: https://doi.org/10.1029/2020GL089289.

[20] Brooks, R. H. 1965. Hydraulic Properties of Porous Media. Colorado State University.

[21] van Genuchten, M. T. 1980. A closed-form equation for predicting the hydraulic conductivity of unsaturated soils. Soil Science Society of America Journal 44(5): 892–898. doi: https://doi.org/10.2136/sssaj1980.03615995004400050002x.

[22] Singh, H., Myshakin, E. M. and Seol, Y. 2020. A novel relative permeability model for gas and water flow in hydrate-bearing sediments with laboratory and field-scale application. Scientific Reports 10(1): 5697, 2020/03/30 2020, doi: 10.1038/s41598-020-62284-5.

[23] Liu, X. and Flemings, P. B. 2011. Capillary effects on hydrate stability in marine sediments. Journal of Geophysical Research: Solid Earth 116(B7). doi: https://doi.org/10.1029/2010JB008143.

[24] Chen, X. and Espinoza, D. N. 2018. Ostwald ripening changes the pore habit and spatial variability of clathrate hydrate. Fuel 214: 614–622, 2018/02/15/ 2018, doi: https://doi.org/10.1016/j.fuel.2017.11.065.

[25] Lei, L., Seol, Y. and Myshakin, E. M. 2019. Methane hydrate film thickening in porous media. Geophysical Research Letters 46(20): 11091–11099. doi: https://doi.org/10.1029/2019GL084450.

[26] Waite, W. F. et al. 2009. Physical properties of hydrate-bearing sediments. Reviews of Geophysics 47(4). doi: https://doi.org/10.1029/2008RG000279.

[27] Clennell, M. B., Hovland, M., Booth, J. S., Henry, P. and Winters, W. J. 1999. Formation of natural gas hydrates in marine sediments: 1. Conceptual model of gas hydrate growth conditioned by host sediment properties. Journal of Geophysical Research: Solid Earth 104(B10): 22985–23003. doi: https://doi.org/10.1029/1999JB900175.

[28] Kleinberg, R. L., Flaum, C., Griffin, D. D., Brewer, P. G., Malby, G. E., Peltzer, E. T. et al. 2003. Deep sea NMR: Methane hydrate growth habit in porous media and its relationship to hydraulic permeability, deposit accumulation, and submarine slope stability. Journal of Geophysical Research: Solid Earth 108(B10). doi: https://doi.org/10.1029/2003JB002389.

[29] Mahabadi, N., Dai, S., Seol, Y. and Jang, J. 2019. Impact of hydrate saturation on water permeability in hydrate-bearing sediments. Journal of Petroleum Science and Engineering 174: 696–703, 2019/03/01/ 2019, doi: https://doi.org/10.1016/j.petrol.2018.11.084.

[30] Priest, J. A., Rees, E. V. L. and Clayton, C. R. I. 2009. Influence of gas hydrate morphology on the seismic velocities of sands. Journal of Geophysical Research: Solid Earth 114(B11). doi: https://doi.org/10.1029/2009JB006284.

[31] Spangenberg, E. 2001. Modeling of the influence of gas hydrate content on the electrical properties of porous sediments. Journal of Geophysical Research: Solid Earth 106(B4): 6535–6548, doi: https://doi.org/10.1029/2000JB900434.

[32] Rajabi, M. S. and Moradi, R. 2023. Fossil fuels reservoirs and extraction. In Reference Module in Earth Systems and Environmental Sciences: Elsevier.

[33] Rackley, S. A. 2017. 18 - Other geological storage options. pp. 471–488. *In*: Carbon Capture and Storage (Second Edition), S. A. Rackley Ed. Boston: Butterworth-Heinemann.

[34] Khalilnezhad, A., Rezvani, H., Abdi, A. and Riazi, M. 2023. Chapter 8 - Hybrid thermal chemical EOR methods. pp. 269–314. *In*: Hemmati-Sarapardeh, A., Alamatsaz, A., Dong, M. and Li, Z. (eds.). Thermal Methods: Gulf Professional Publishing.

[35] El-hoshoudy, A. N. and Desouky, S. 2018. CO_2 miscible flooding for enhanced oil recovery. *In*: Ramesh, K. A. (ed.). Carbon Capture, Utilization and Sequestration. Rijeka: IntechOpen, Chapter 5.

[36] Khoshsima, A., Sedighi, M. and Mohammadi, M. 2023. Chapter 8 - Enhanced oil recovery by water alternating gas injection. pp. 295–316. *In*: Li, Z., Husein, M. M. and Hemmati-Sarapardeh, A. (eds.). Gas Injection Methods, Gulf Professional Publishing.

[37] Liu, S.-Y., Ren, B., Li, H.-Y., Yang, Y.-Z., Wang, Z.-Q., Wang, B. et al. 2022. CO_2 storage with enhanced gas recovery (CSEGR): A review of experimental and numerical studies. Petroleum Science 19(2): 594–607, 2022/04/01/ 2022, doi: https://doi.org/10.1016/j.petsci.2021.12.009.

[38] Liao, H., Pan, W., He, Y., Fang, X. and Zhang, Y. 2023. Study on the mechanism of CO_2 injection to improve tight sandstone gas recovery. Energy Reports 9: 645–656, 2023/12/01/ 2023, doi: https://doi.org/10.1016/j.egyr.2022.11.210.

[39] Iddphonce, R., Wang, J. and Zhao, L. 2020. Review of CO_2 injection techniques for enhanced shale gas recovery: Prospect and challenges. Journal of Natural Gas Science and Engineering 77: 103240, 2020/05/01/ 2020, doi: https://doi.org/10.1016/j.jngse.2020.103240.

[40] Fan, C., Yang, L., Sun, H., Luo, M., Zhou, L., Yang, Z. et al. 2023. Recent advances and perspectives of CO_2-enhanced coalbed methane: experimental, modeling, and

technological development. Energy & Fuels 37(5): 3371–3412, 2023/03/02 2023, doi: 10.1021/acs.energyfuels.2c03823.

[41] Kumar, S., Foroozesh, J., Edlmann, K., Rezk, M. G. and Lim, C. Y. 2020. A comprehensive review of value-added CO_2 sequestration in subsurface saline aquifers. Journal of Natural Gas Science and Engineering 81: 103437, 2020/09/01/ 2020, doi: https://doi.org/10.1016/j.jngse.2020.103437.

[42] Michael, K., Golab, A., Shulakova, V., Ennis-King, J., Allinson, G., Sharma, S. et al. 2010. Geological storage of CO_2 in saline aquifers—A review of the experience from existing storage operations. International Journal of Greenhouse Gas Control 4(4): 659–667, 2010/07/01/ 2010, doi: https://doi.org/10.1016/j.ijggc.2009.12.011.

[43] Cui, G., Hu, Z., Ning, F., Jiang, S. and Wang, R. 2023. A review of salt precipitation during CO_2 injection into saline aquifers and its potential impact on carbon sequestration projects in China. Fuel 334: 126615, 2023/02/15/ 2023, doi: https://doi.org/10.1016/j.fuel.2022.126615.

[44] Ezekiel, J., Ebigbo, A., Adams, B. M. and Saar, M. O. 2020. Combining natural gas recovery and CO_2-based geothermal energy extraction for electric power generation. Applied Energy 269: 115012, 2020/07/01/ 2020, doi: https://doi.org/10.1016/j.apenergy.2020.115012.

[45] Wu, Y. and Li, P. 2020. The potential of coupled carbon storage and geothermal extraction in a CO_2-enhanced geothermal system: a review. Geothermal Energy 8(1): 19, 2020/06/15 2020, doi: 10.1186/s40517-020-00173-w.

[46] Selley, R. 2005. Mineralogy and Classification.

[47] Kang, S.-P. and Lee, J.-W. 2010. Kinetic behaviors of CO_2 hydrates in porous media and effect of kinetic promoter on the formation kinetics. Chemical Engineering Science 65(5): 1840–1845, 2010/03/01/ 2010, doi: https://doi.org/10.1016/j.ces.2009.11.027.

[48] Song, Y., Wang, S., Jiang, L., Zhang, Y. and Yang, M. 2016. Hydrate phase equilibrium for CH_4-CO_2-H_2O system in porous media. The Canadian Journal of Chemical Engineering 94(8): 1592–1598, doi: https://doi.org/10.1002/cjce.22529.

[49] Song, Y., Wang, X., Yang, M., Jiang, L., Liu, Y., Dou, B. et al. 2013. Study of selected factors affecting hydrate-based carbon dioxide separation from simulated fuel gas in porous media. Energy & Fuels 27(6): 3341–3348, 2013/06/20 2013, doi: 10.1021/ef400257a.

[50] Yang, M., Song, Y., Liu, Y., Jiang, L. and Zhao, Y. 2016. Thermodynamic Characters of N_2/CO_2 hydrates in marine sediment. In The 26th International Ocean and Polar Engineering Conference, vol. All Days, ISOPE-I-16-220.

[51] Kumar, A., Sakpal, T., Linga, P. and Kumar, R. 2013. Influence of contact medium and surfactants on carbon dioxide clathrate hydrate kinetics. Fuel 105: 664–671, 2013/03/01/ 2013, doi: https://doi.org/10.1016/j.fuel.2012.10.031.

[52] Kumar, A., Sakpal, T., Linga, P. and Kumar, R. 2015. Enhanced carbon dioxide hydrate formation kinetics in a fixed bed reactor filled with metallic packing. Chemical Engineering Science 122: 78–85, 2015/01/27/ 2015, doi: https://doi.org/10.1016/j.ces.2014.09.019.

[53] Nambiar, A., Babu, P. and Linga, P. 2015. CO_2 capture using the clathrate hydrate process employing cellulose foam as a porous media. Canadian Journal of Chemistry 93(8): 808–814. doi: 10.1139/cjc-2014-0547.

[54] Yang, M., Song, Y., Ruan, X., Liu, Y., Zhao, J. and Li, Q. 2012. Characteristics of CO_2 hydrate formation and dissociation in glass beads and silica gel. Energies 5(4): 925–937 [Online]. Available: https://www.mdpi.com/1996-1073/5/4/925.

[55] Kumar, A., Palodkar, A. V., Gautam, R., Choudhary N., Veluswamy, H. P. and Kumar, S. 2022. Role of salinity in clathrate hydrate based processes. Journal of Natural Gas Science and Engineering 108: 104811, 2022/12/01/ 2022, doi: https://doi.org/10.1016/j.jngse.2022.104811.

[56] Rehman, A. N., Bavoh, C. B., Pendyala, R. and Lal, B. 2021. Research advances, maturation, and challenges of hydrate-based CO_2 sequestration in porous media. ACS Sustainable Chemistry & Engineering 9(45): 15075–15108, 2021/11/15 2021, doi: 10.1021/acssuschemeng.1c05423.
[57] Mekala, P., Busch, M., Mech, D., Patel, R. S. and Sangwai, J. S. 2014. Effect of silica sand size on the formation kinetics of CO_2 hydrate in porous media in the presence of pure water and seawater relevant for CO_2 sequestration. Journal of Petroleum Science and Engineering 122: 1–9, 2014/10/01/ 2014, doi: https://doi.org/10.1016/j.petrol.2014.08.017.
[58] Yang, S. H. B., Babu, P., Chua, S. F. S. and Linga, P. 2016. Carbon dioxide hydrate kinetics in porous media with and without salts. Applied Energy 162: 1131–1140, 2016/01/15/ 2016, doi: https://doi.org/10.1016/j.apenergy.2014.11.052.
[59] Bhattacharjee, G., Kumar, A., Sakpal, T. and Kumar, R. 2015. Carbon dioxide sequestration: influence of porous media on hydrate formation kinetics. ACS Sustainable Chemistry & Engineering 3(6): 1205–1214, 2015/06/01 2015, doi: 10.1021/acssuschemeng.5b00171.
[60] Lee, J.-w., Chun, M.-K., Lee, K.-M., Kim, Y.-J. and Lee, H. 2002. Phase equilibria and kinetic behavior of CO_2 hydrate in electrolyte and porous media solutions: application to ocean sequestration of CO_2. Korean Journal of Chemical Engineering 19(4): 673–678, 2002/07/01 2002, doi: 10.1007/BF02699316.
[61] Arora, A., Kumar, A., Bhattacharjee, G., Kumar, P. and Balomajumder, C. 2016. Effect of different fixed bed media on the performance of sodium dodecyl sulfate for hydrate based CO_2 capture. Materials & Design 90: 1186–1191, 2016/01/15/ 2016, doi: https://doi.org/10.1016/j.matdes.2015.06.049.
[62] Sun, D. and Englezos, P. 2014. Storage of CO_2 in a partially water saturated porous medium at gas hydrate formation conditions. International Journal of Greenhouse Gas Control 25: 1–8, 2014/06/01/ 2014, doi: https://doi.org/10.1016/j.ijggc.2014.03.008.
[63] Zhang, X., Li, J., Wu, Q., Wang, C. and Nan, J. 2015. Experimental study on the effect of pore size on carbon dioxide hydrate formation and storage in porous media. Journal of Natural Gas Science and Engineering 25: 297–302, 2015/07/01/ 2015, doi: https://doi.org/10.1016/j.jngse.2015.05.014.
[64] Yang, M., Song, Y., Jiang, L., Zhu, N., Liu, Y., Zhao, Y. et al. 2013. CO_2 hydrate formation and dissociation in cooled porous media: a potential technology for CO_2 capture and storage. Environmental Science & Technology 47(17): 9739–9746, 2013/09/03 2013, doi: 10.1021/es401536w.
[65] Kang, S.-P., Seo, Y. and Jang, W. 2009. Kinetics of methane and carbon dioxide hydrate formation in silica gel pores. Energy & Fuels 23(7): 3711–3715, 2009/07/16 2009, doi: 10.1021/ef900256f.
[66] Seo, Y., Lee, H. and Uchida, T. 2002. Methane and carbon dioxide hydrate phase behavior in small porous silica gels: three-phase equilibrium determination and thermodynamic modeling. Langmuir 18(24): 9164–9170, 2002/11/01 2002, doi: 10.1021/la0257844.
[67] Sinehbaghizadeh, S., Saptoro, A. and Mohammadi, A. H. 2022. CO_2 hydrate properties and applications: A state of the art. Progress in Energy and Combustion Science 93: 101026, 2022/11/01/ 2022, doi: https://doi.org/10.1016/j.pecs.2022.101026.
[68] Kang, S.-P., Lee, J.-W. and Ryu, H.-J. 2008. Phase behavior of methane and carbon dioxide hydrates in meso- and macro-sized porous media. Fluid Phase Equilibria 274(1): 68–72, 2008/12/25/ 2008, doi: https://doi.org/10.1016/j.fluid.2008.09.003.
[69] Zhang, Y., Li, X., Chen, Z., Cai, J., Xu, C. and Li, G. 2017. Formation behaviors of CO_2 hydrate in Kaoline and Bentonite clays with partially water saturated. Energy Procedia 143: 547–552, 2017/12/01/ 2017, doi: https://doi.org/10.1016/j.egypro.2017.12.724.
[70] Zheng, J.-n., Yang, L., Ma, S., Zhao, Y. and Yang, M. 2020. Quantitative analysis of CO_2 hydrate formation in porous media by proton NMR. AIChE Journal 66(2): e16820. doi: https://doi.org/10.1002/aic.16820.

[71] Babu, P., Kumar, R. and Linga, P. 2013. Pre-combustion capture of carbon dioxide in a fixed bed reactor using the clathrate hydrate process. Energy 50: 364–373, 2013/02/01/ 2013, doi: https://doi.org/10.1016/j.energy.2012.10.046.
[72] Yang, M., Liu, W., Song, Y., Ruan, X., Wang, X., Zhao, J. et al. 2013. Effects of additive mixture (THF/SDS) on the thermodynamic and kinetic properties of CO_2/H_2 hydrate in porous media. Industrial & Engineering Chemistry Research 52(13): 4911–4918, 2013/04/03 2013, doi: 10.1021/ie303280e.
[73] Zhang, X., Li, J., Wu, Q., Wang, Y., Wang, J. and Li, Y. 2019. Effect of initial pressure on the formation of carbon dioxide hydrate in frozen quartz sand. Energy & Fuels 33(11): 11346–11352, 2019/11/21 2019, doi: 10.1021/acs.energyfuels.9b01693.
[74] Zheng, J., Zhang, B.-Y., Wu, Q. and Linga, P. 2018. Kinetic evaluation of cyclopentane as a promoter for CO_2 capture via a clathrate process employing different contact modes. ACS Sustainable Chemistry & Engineering 6(9): 11913–11921, 2018/09/04 2018, doi: 10.1021/acssuschemeng.8b02187.
[75] Lee, K., Lee, S.-H. and Lee, W. 2013. Stochastic nature of carbon dioxide hydrate induction times in Na-montmorillonite and marine sediment suspensions. International Journal of Greenhouse Gas Control 14: 15–24, 2013/05/01/ 2013, doi: https://doi.org/10.1016/j.ijggc.2013.01.001.
[76] Zheng, J.-n. and Yang, M. 2019. Phase equilibrium data of CO_2–MCP hydrates and CO_2 gas uptake comparisons with CO_2–CP hydrates and CO_2–C_3H_8 hydrates. Journal of Chemical & Engineering Data 64(1): 372–379, 2019/01/10 2019, doi: 10.1021/acs.jced.8b00893.
[77] Abu Hassan, M. H., Sher, F., Zarren, G., Suleiman, N., Tahir, A. A. and Snape, C. E. 2020. Kinetic and thermodynamic evaluation of effective combined promoters for CO_2 hydrate formation. Journal of Natural Gas Science and Engineering 78: 103313, 2020/06/01/ 2020, doi: https://doi.org/10.1016/j.jngse.2020.103313.
[78] Seo, Y.-T., Moudrakovski, I. L., Ripmeester, J. A., Lee, J.-w. and Lee, H. 2005. Efficient recovery of CO_2 from flue gas by clathrate hydrate formation in porous silica gels. Environmental Science & Technology 39(7): 2315–2319, 2005/04/01 2005, doi: 10.1021/es049269z.
[79] Kang, S.-P., Lee, J. and Seo, Y. 2013. Pre-combustion capture of CO_2 by gas hydrate formation in silica gel pore structure. Chemical Engineering Journal 218: 126–132, 2013/02/15/ 2013, doi: https://doi.org/10.1016/j.cej.2012.11.131.
[80] Kim, D., Ahn, Y.-H. and Lee, H. 2015. Phase equilibria of CO_2 and CH_4 hydrates in intergranular meso/macro pores of MIL-53 metal organic framework. Journal of Chemical & Engineering Data 60(7): 2178–2185, 2015/07/09 2015, doi: 10.1021/acs.jced.5b00322.
[81] Kang, S.-P., Seo, Y. and Jang, W. 2009. Gas hydrate process for recovery of CO_2 from fuel gas. Chemical Engineering Transactions 17: 1449–1454.
[82] Kim, D., Ahn, Y.-H., Kim, S.-J., Lee, J. Y., Lee, J., Seo, Y.-j. et al. 2015. Gas hydrate in crystalline-swelled clay: the effect of pore dimension on hydrate formation and phase equilibria. The Journal of Physical Chemistry C 119(38): 22148–22153, 2015/09/24 2015, doi: 10.1021/acs.jpcc.5b03229.
[83] Mu, L. and Cui, Q. 2019. Experimental study on the dissociation equilibrium of (CH_4 + CO_2) hydrates in the (Quartz Sands + NaCl Solution) system. Journal of Chemical & Engineering Data 64(12): 6041–6048, 2019/12/12 2019, doi: 10.1021/acs.jced.9b00859.
[84] Anderson, R., Llamedo, M., Tohidi, B. and Burgass, R. W. 2003. Experimental measurement of methane and carbon dioxide clathrate hydrate equilibria in mesoporous silica. The Journal of Physical Chemistry B 107(15): 3507–3514, 2003/04/01 2003, doi: 10.1021/jp0263370.

[85] Park, T., Kyung, D. and Lee, W. 2014. Effect of organic matter on CO_2 hydrate phase equilibrium in phyllosilicate suspensions. Environmental Science & Technology 48(12): 6597–6603, 2014/06/17 2014, doi: 10.1021/es405099z.

[86] Li, S.-L., Ma, Q.-L., Sun, C.-Y., Chen, L.-T., Liu, B., Feng, X.-J. et al. 2013. A fractal approach on modeling gas hydrate phase equilibria in porous media. Fluid Phase Equilibria 356: 277–283, 2013/10/25/ 2013, doi: https://doi.org/10.1016/j.fluid.2013.07.047.

[87] Wang, J., Zhao, J., Zhang, Y., Wang, D., Li, Y. and Song, Y. 2016. Analysis of the effect of particle size on permeability in hydrate-bearing porous media using pore network models combined with CT. Fuel 163: 34–40, 2016/01/01/ 2016, doi: https://doi.org/10.1016/j.fuel.2015.09.044.

[88] Yang, M., Song, Y., Zhao, Y., Liu, Y., Jiang, L. and Li, Q. 2011. MRI measurements of CO_2 hydrate dissociation rate in a porous medium. Magnetic Resonance Imaging 29(7): 1007–1013, 2011/09/01/ 2011, doi: https://doi.org/10.1016/j.mri.2011.04.008.

[89] Hosseini Zadeh, A., Kim, I. and Kim, S. 2020. Experimental study on the characteristics of formation and dissociation of CO_2 hydrates in porous media. E3S Web Conf. 205: 02004 [Online]. Available: https://doi.org/10.1051/e3sconf/202020502004.

[90] Sun, X., Qin, X., Lu, H., Wang, J., Xu, J. and Ning, Z. 2021. Gas hydrate *in-situ* formation and dissociation in clayey-silt sediments: An investigation by low-field NMR. Energy Exploration & Exploitation 39(1): 256–272. doi: 10.1177/0144598720974159.

[91] Li, X.-Y., Wang, Y., Li, X.-S., Zhang, Y. and Chen, Z.-Y. 2019. Experimental study of methane hydrate dissociation in porous media with different thermal conductivities. International Journal of Heat and Mass Transfer 144: 118528, 2019/12/01/ 2019, doi: https://doi.org/10.1016/j.ijheatmasstransfer.2019.118528.

[92] Li, X.-Y., Li, X.-S., Wang, Y., Li, G., Zhang, Y., Hu, H.-Q. et al. 2021. Influence of particle size on the heat and mass transfer characteristics of methane hydrate formation and decomposition in porous media. Energy & Fuels 35(3): 2153–2164, 2021/02/04 2021, doi: 10.1021/acs.energyfuels.0c03812.

[93] Yang, M., Sun, H., Chen, B. and Song, Y. 2019. Effects of water-gas two-phase flow on methane hydrate dissociation in porous media. Fuel 255: 115637, 2019/11/01/ 2019, doi: https://doi.org/10.1016/j.fuel.2019.115637.

[94] Guo, P., Pan, Y.-K., Li, L.-L. and Tang, B. 2017. Molecular dynamics simulation of decomposition and thermal conductivity of methane hydrate in porous media. Chinese Physics B 26(7): 073101, 2017/06/01 2017, doi: 10.1088/1674-1056/26/7/073101.

[95] Chen, X. and Espinoza, D. N. 2018. Surface area controls gas hydrate dissociation kinetics in porous media. Fuel 234: 358–363, 2018/12/15/ 2018, doi: https://doi.org/10.1016/j.fuel.2018.07.030.

[96] Muhammad Saad, K., Bavoh, B. C., Bhajan, L. and Mohamad Azmi, B. 2018. Kinetic assessment of tetramethyl ammonium hydroxide (ionic liquid) for carbon dioxide, methane and binary mix gas hydrates. *In*: Mohammed Muzibur, R. (ed.). Recent Advances in Ionic Liquids. Rijeka: IntechOpen, 2018, p. Ch. 9.

[97] Ge, B.-B., Li, X.-Y., Zhong, D.-L. and Lu, Y.-Y. 2022. Investigation of natural gas storage and transportation by gas hydrate formation in the presence of bio-surfactant sulfonated lignin. Energy 244: 122665, 2022/04/01/ 2022, doi: https://doi.org/10.1016/j.energy.2021.122665.

[98] Yi, J., Zhong, D.-L., Yan, J. and Lu, Y.-Y. 2019. Impacts of the surfactant sulfonated lignin on hydrate based CO_2 capture from a CO_2/CH_4 gas mixture. Energy 171: 61–68, 2019/03/15/ 2019, doi: https://doi.org/10.1016/j.energy.2019.01.007.

[99] Sahu, C., Sircar, A., Sangwai, J. S. and Kumar, R. 2022. Effect of methylamine, amylamine, and decylamine on the formation and dissociation kinetics of CO_2 hydrate relevant for carbon dioxide sequestration. Industrial & Engineering Chemistry Research 61(7): 2672–2684, 2022/02/23 2022, doi: 10.1021/acs.iecr.1c04074.

[100] Rackley, S. A. 2017. 20 - Ocean storage. pp. 517–541. *In*: Carbon Capture and Storage (Second Edition), S. A. Rackley Ed. Boston: Butterworth-Heinemann.
[101] Boswell, R. 2009. Is gas hydrate energy within reach? Science 325(5943): 957–958. doi: doi:10.1126/science.1175074.
[102] Holder, G. D., Kamath, a. V. A. and Godbole, S. P. 1984. The potential of natural gas hydrates as an energy resource. Annual Review of Energy 9(1): 427–445. doi: 10.1146/annurev.eg.09.110184.002235.
[103] Zheng, J., Chong, Z. R., Qureshi, M. F. and Linga, P. 2020. Carbon dioxide sequestration via gas hydrates: a potential pathway toward decarbonization. Energy & Fuels 34(9): 10529–10546, doi: 10.1021/acs.energyfuels.0c02309.
[104] Melnikov, V. and Drozdov, D. 2006. Distribution of permafrost in Russia. pp. 67–80. *In*: Advances in the Geological Storage of Carbon Dioxide: Springer.
[105] Le Nindre, Y.-M., Allier, D., Duchkov, A., Altunina, L. K., Shvartsev, S., Zhelezniak, M. et al. 2011. Storing CO_2 underneath the Siberian Permafrost: A win-win solution for long-term trapping of CO_2 and heavy oil upgrading. Energy Procedia 4: 5414–5421, 2011/01/01/ 2011, doi: https://doi.org/10.1016/j.egypro.2011.02.526.
[106] Duchkov, A. D., Permyakov, M. E., Sokolova, L. S., Ayunov, D. E. and Trofimuk. 2011. Assessment of possibility for the carbon dioxide storage in West Siberian permafrost.
[107] Birchwood, R., Noeth, S. and Jones, E. 2008. Safe drilling in gasthydrate prone sediments: findings from the 2005 drilling campaign of The Gulf of Mexico Gas Hydrates Joint Industry Project (JIP). Natural Gas & Oil 304: 285–4541.
[108] Brewer, P. G., Friederich, G., Peltzer, E. T. and Orr, F. M. 1999. Direct experiments on the ocean disposal of fossil fuel CO_2. Science 284(5416): 943–945. doi: doi:10.1126/science.284.5416.943.
[109] Lee, S., Liang, L., Riestenberg, D., West, O. R., Tsouris, C. and Adams, E. 2003. CO_2 hydrate composite for ocean carbon sequestration. Environmental Science & Technology 37(16): 3701–3708, 2003/08/01 2003, doi: 10.1021/es026301l.
[110] Tsouris, C., Brewer, P., Peltzer, E., Walz, P., Riestenberg, D., Liang, L. et al. 2004. Hydrate composite particles for ocean carbon sequestration: field verification. Environmental Science & Technology 38(8): 2470–2475, 2004/04/01 2004, doi: 10.1021/es034990a.
[111] Riestenberg, D. E., Tsouris, C., Brewer, P. G., Peltzer, E. T., Walz, P., Chow, A. C. et al. 2005. Field studies on the formation of sinking CO_2 particles for ocean carbon sequestration: effects of injector geometry on particle density and dissolution rate and model simulation of plume behavior. Environmental Science & Technology 39(18): 7287–7293, 2005/09/01 2005, doi: 10.1021/es050125+.
[112] Tsouris, C., Szymcek, P., Taboada-Serrano, P., McCallum, S. D., Brewer, P., Peltzer, E. et al. 2007. Scaled-up ocean injection of CO_2–hydrate composite particles. Energy & Fuels 21(6): 3300–3309, 2007/11/01 2007, doi: 10.1021/ef070197h.
[113] Tsouris, C., McCallum, S., Aaron, D., Riestenberg, D., Gabitto, J., Chow, A. et al. 2007. Scale-up of a continuous-jet hydrate reactor for CO_2 ocean sequestration. AIChE Journal 53(4): 1017–1027. doi: https://doi.org/10.1002/aic.11117.
[114] Szymcek, P., McCallum, S. D., Taboada-Serrano, P. and Tsouris, C. 2008. A pilot-scale continuous-jet hydrate reactor. Chemical Engineering Journal 135(1): 71–77, 2008/01/15/ 2008, doi: https://doi.org/10.1016/j.cej.2007.03.029.
[115] Warzinski, R. P., Riestenberg, D. E., Gabitto, J., Haljasmaa, I. V., Lynn, R. J. and Tsouris, C. 2008. Formation and behavior of composite CO_2 hydrate particles in a high-pressure water tunnel facility. Chemical Engineering Science 63(12): 3235–3248, 2008/06/01/ 2008, doi: https://doi.org/10.1016/j.ces.2008.03.005.
[116] GHG, I. 2004. Gas hydrates for deep ocean storage of CO_2. IEA Greenhouse Gas R&D Programme Report PH4/26.
[117] Rehder, G., Kirby, S. H., Durham, W. B., Stern, L. A., Peltzer, E. T., Pinkston, J. et al. 2004. Dissolution rates of pure methane hydrate and carbon-dioxide hydrate

in undersaturated seawater at 1000-m depth. Geochimica et Cosmochimica Acta 68(2): 285–292, 2004/01/15/ 2004, doi: https://doi.org/10.1016/j.gca.2003.07.001.
[118] Tamburri, M. N., Peltzer, E. T., Friederich, G. E., Aya, I., Yamane, K. and Brewer, P. G. 2000. A field study of the effects of CO_2 ocean disposal on mobile deep-sea animals. Marine Chemistry 72(2): 95–101, 2000/12/01/ 2000, doi: https://doi.org/10.1016/S0304-4203(00)00075-X.
[119] Barry, J. P., Buck, K. R., Lovera, C. F., Kuhnz, L., Whaling, P. J., Peltzer, E. T. et al. 2004. Effects of direct ocean CO_2 injection on deep-sea meiofauna. Journal of Oceanography 60(4): 759–766, 2004/08/01 2004, doi: 10.1007/s10872-004-5768-8.
[120] Goel, N. 2006. *In situ* methane hydrate dissociation with carbon dioxide sequestration: Current knowledge and issues. Journal of Petroleum Science and Engineering 51(3): 169–184, 2006/05/16/ 2006, doi: https://doi.org/10.1016/j.petrol.2006.01.005.
[121] Ebinuma, T. 1993. Method for Dumping and Disposing of Carbon Dioxide Gas and Apparatus Therefor. ed: Google Patents.
[122] Ohgaki, K., Takano, K., Sangawa, H., Matsubara, T. and Nakano, S. 1996. Methane exploitation by carbon dioxide from gas hydrates – phase equilibria for CO_2-CH_4 mixed hydrate system. Journal of Chemical Engineering of Japan 29(3): 478–483. doi: 10.1252/jcej.29.478.
[123] Wang, Y., Lang, X., Fan, S., Wang, S., Yu, C. and Li, G. 2021. Review on enhanced technology of natural gas hydrate recovery by carbon dioxide replacement. Energy & Fuels 35(5): 3659–3674, 2021/03/04 2021, doi: 10.1021/acs.energyfuels.0c04138.
[124] McConnell, D. R., Zhang, Z. and Boswell, R. 2012. Review of progress in evaluating gas hydrate drilling hazards. Marine and Petroleum Geology 34(1): 209–223, 2012/06/01/ 2012, doi: https://doi.org/10.1016/j.marpetgeo.2012.02.010.
[125] Ota, M., Saito, T., Aida, T., Watanabe, M., Sato, Y., Smith Jr., R. L. et al. 2007. Macro and microscopic CH_4–CO_2 replacement in CH_4 hydrate under pressurized CO_2. AIChE Journal 53(10): 2715–2721. doi: https://doi.org/10.1002/aic.11294.
[126] Kim, H. C., Bishnoi, P. R., Heidemann, R. A. and Rizvi, S. S. H. 1987. Kinetics of methane hydrate decomposition. Chemical Engineering Science 42(7): 1645–1653, 1987/01/01/ 1987, doi: https://doi.org/10.1016/0009-2509(87)80169-0.
[127] Wang, T., Zhang, L., Sun, L., Zhou, R., Dong, B., Yang, L. et al. 2021. Methane recovery and carbon dioxide storage from gas hydrates in fine marine sediments by using CH_4/CO_2 replacement. Chemical Engineering Journal 425: 131562, 2021/12/01/ 2021, doi: https://doi.org/10.1016/j.cej.2021.131562.
[128] Zhao, J., Zhang, L., Chen, X., Fu, Z., Liu, Y. and Song, Y. 2015. Experimental study of conditions for methane hydrate productivity by the CO_2 swap method. Energy & Fuels 29(11): 6887–6895, 2015/11/19 2015, doi: 10.1021/acs.energyfuels.5b00913.
[129] Pandey, J. S., Karantonidis, C., Karcz, A. P. and von Solms, N. 2020. Enhanced CH_4-CO_2 hydrate swapping in the presence of low dosage methanol. Energies 13(20): 5238 [Online]. Available: https://www.mdpi.com/1996-1073/13/20/5238.
[130] Brewer, P. G., Peltzer, E. T., Walz, P. M., Coward, E. K., Stern, L. A., Kirby, S. H. et al. 2014. Deep-sea field test of the CH_4 hydrate to CO_2 hydrate spontaneous conversion hypothesis. Energy & Fuels 28(11): 7061–7069, 2014/11/20 2014, doi: 10.1021/ef501430h.
[131] Boswell, R., Schoderbek, D., Collett, T. S., Ohtsuki, S., White, M. and Anderson, B. J. 2017. The Iġnik Sikumi field experiment, Alaska North Slope: Design, operations, and implications for CO_2–CH_4 exchange in gas hydrate reservoirs. Energy & Fuels 31(1): 140–153, 2017/01/19 2017, doi: 10.1021/acs.energyfuels.6b01909.
[132] Schoderbek, D. and Boswell, R. 2011. Iġnik Sikumi# 1, gas hydrate test well, successfully installed on the Alaska North Slope. Natural Gas & Oil 304: 285–4541.
[133] Kvamme, B., Graue, A., Buanes, T., Kuznetsova, T. and Ersland, G. 2007. Storage of CO_2 in natural gas hydrate reservoirs and the effect of hydrate as an extra sealing in cold

aquifers. International Journal of Greenhouse Gas Control 1(2): 236–246, 2007/04/01/ 2007, doi: https://doi.org/10.1016/S1750-5836(06)00002-8.

[134] Birkedal, K. A., Ersland, G., Husebo, J., Kvamme, B. and Graue, A. 2010. Geomechanical stability during CH_4 production from hydrates - depressurization or CO_2 sequestration with CO_2-CH_4 exchange. In 44th U.S. Rock Mechanics Symposium and 5th U.S.-Canada Rock Mechanics Symposium, vol. All Days, ARMA-10-321.

[135] Schoderbek, D., Farrell, H., Howard, J., Raterman, K., Silpngarmlert, S., Martin, K. et al. 2013. ConocoPhillips Gas Hydrate Production Test. United States [Online]. Available: https://www.osti.gov/biblio/1123878.

[136] Koh, D.-Y., Kang, H., Lee, J.-W., Park, Y., Kim, S.-J., Lee, J. et al. 2016. Energy-efficient natural gas hydrate production using gas exchange. Applied Energy 162: 114–130, 2016/01/15/ 2016, doi: https://doi.org/10.1016/j.apenergy.2015.10.082.

CHAPTER 7
Risk Analysis in Carbon Capture and Storage

*Grace Amabel Tabaaza** and *Bhajan Lal*

7.1 Introduction

Risk analysis is a critical component in the implementation of Carbon Capture and Storage (CCS) projects, involving the identification, assessment, and mitigation of potential hazards associated with CCS. The key risks in CCS include leakage of stored CO_2, environmental impacts, health and safety risks, data insufficiency, and regulatory challenges. Environmental risk assessment is crucial, focusing on evaluating potential impacts such as CO_2 leakage on ecosystems and human health. Risks are categorized into short-term and long-term, requiring qualitative and quantitative risk assessment methods. Organizations like the Environment Agency and Tetra Tech actively conduct risk assessments.

Risk analysis is an ongoing process throughout the CCS project lifecycle, with ongoing research and development efforts aimed at improving safety and effectiveness. To mitigate risks, comprehensive risk management plans, a risk-based approach, rigorous risk identification and assessment, safety technology development, collaboration with experts, and personnel training are crucial. These strategies and measures are tailored to the unique challenges of Carbon Capture, Utilization, and Storage (CCUS) facilities, ensuring secure and effective CCS technology implementation to mitigate climate impacts.

Chemical Engineering Department, Center for Carbon Capture, Storage and Utilization (CCUS), Institute of Sustainable Energy, Universiti Teknologi PETRONAS, Bandar Seri Iskandar, Perak, Malaysia.
* Corresponding author: tabaazagrace@gmail.com

7.2 Risk Assessment in Carbon Capture and Storage

Risk analysis plays a crucial role in the implementation of carbon capture and storage (CCS) projects. Risk analysis in CCS is an assessment that involves identifying, analyzing, and evaluating potential risks associated with CCS projects. It aims to understand the likelihood and consequences of various hazards and develop strategies to mitigate them [1]. Potential risks associated with carbon capture and storage (CCS) include:

Leakage: The largest and most obvious risk of CCS is the potential for leakage. When compressed carbon is stored in underground reservoirs, there is a small risk of abrupt or gradual leakage, which could release stored carbon dioxide (CO_2) into the atmosphere [1]. CCS projects therefore have the potential to cause negative environmental impacts.

Underground storage poses risks such as potential leakage, contamination of drinking water, and stimulation of seismic activity [2]. Injecting large amounts of captured CO_2 into underground formations could lead to unforeseen consequences and long-term damage to the environment [3].

In addition, the buildout of CCS infrastructure presents health and safety risks. Marginalized communities already burdened by industrial hazards may be disproportionately affected by CCS projects [2]. The capture, transport, and injection of CO_2 involve potential short-term accidents that can pose risks to workers and nearby communities [4].

There is still insufficient verifiable data to demonstrate that the effectiveness of CCS technology is a key impediment to CCS investments. The lack of comprehensive data on project costs, technical performance, and the long-term permanence of CO_2 storage hinders the understanding of the actual deployment and mitigation costs of CCS [5].

It is important to note that the above risks can be mitigated through proper risk assessment, monitoring, and regulatory oversight. Ongoing research and development efforts aim to address these risks and improve the safety and effectiveness of CCS technologies.

7.3 Environmental Risk Assessment

Environmental risk assessment focuses on evaluating the potential environmental impacts of CCS, such as the leakage of stored carbon dioxide (CO_2) and its effects on ecosystems and human health. It helps in identifying and addressing regulatory issues and ensuring the safe implementation of CCS [5–6].

Risks in regard to CCS are divided into short-term and long-term risks. Risk analysis for CCS should consider both short-term potential accidents, such as those related to capture, transport, or injection of CO_2, as well as long-term risks associated with CO_2 storage [4]. Then, various qualitative and quantitative risk assessment methods can easily be employed for CCS projects. These methods integrate scientific, engineering, and statistical approaches to evaluate risks and inform decision-making [2].

Organizations like the Environment Agency and Tetra Tech are actively involved in conducting risk assessments for CCS projects. Their aim is to determine the risks associated with carbon capture and storage and develop strategies to mitigate them [3].

It is important to note that risk analysis and assessment are regarded as ongoing processes throughout the lifecycle of CCS projects. This ensures the safe and effective implementation of CCS technologies to mitigate climate impacts.

7.4 Risk Mitigation Strategies and Safety Measures

Effective risk mitigation strategies and safety measures play a pivotal role in guaranteeing the secure functioning of carbon capture, utilization, and storage (CCUS) facilities. Various risk mitigation strategies and safety measures tailored to CCUS a listed here:

Craft a Comprehensive Risk Management Plan: The creation of a robust risk management plan stands as a cornerstone for recognizing, evaluating, and mitigating risks inherent to CCUS endeavours. This plan should encompass a thorough comprehension of the series of events that could precipitate major setbacks within CCUS projects [1].

Implement a Risk-Based Approach: Employing a risk-based approach empowers organizations to allocate their resources to areas, activities, and services posing higher risks to the company, its personnel, and customers. This approach facilitates the development of focused, concise Health, Safety, and Environmental (HSE) standards that are easily comprehensible and amenable to monitoring for efficacy and compliance [7, 8].

Identify and Assess Potential Risks: Risk assessment constitutes the overarching process of pinpointing, scrutinizing, and evaluating potential hazards. Within CCUS initiatives, it is imperative to meticulously analyze and assess health, safety, and environmental risks.

Promote the Development of Safety Technologies: Advancing safety technologies holds significant importance in expediting the adoption of CCUS technology. Innovations in safety technology can enhance the overall safety profile of CCUS facilities.

Forge Collaborative Partnerships with Experts: Collaborating with domain experts in the realm of CCUS can be instrumental in identifying and mitigating risks associated with CCUS projects. Their specialized knowledge and insights can contribute to more effective risk management.

Conduct Comprehensive Personnel Training: Thorough training of personnel is essential to ensure the secure operation of CCUS facilities. Personnel should receive training in safety protocols, emergency response procedures, and risk mitigation strategies.

In summary, the formulation of a robust risk management plan, the adoption of a risk-based approach, the systematic identification and evaluation of potential risks, the advancement of safety technologies, partnering with domain experts, and the meticulous training of personnel all serve as critical risk mitigation strategies and safety measures tailored to the unique challenges of CCUS.

7.5 The Role of CCUS in Mitigating Climate Change

Climate change represents one of the most pressing challenges of our time. The increase in greenhouse gas emissions, primarily carbon dioxide (CO_2), is driving global temperatures upward and causing severe environmental and societal impacts. To combat this crisis, we need innovative solutions that go beyond reducing emissions—we must actively remove CO_2 from the atmosphere. Carbon capture, utilization, and storage (CCUS) is emerging as a pivotal technology in the fight against climate change. This chapter explores how CCUS can significantly contribute to mitigating climate change by capturing, utilizing, and storing CO_2 emissions from industrial processes.

Unlocking Decarbonization: CCUS stands as a vital tool in unlocking the full potential of decarbonization efforts. While transitioning to renewable energy sources and improving energy efficiency are essential steps towards reducing emissions, some sectors, such as heavy industry and certain types of transportation, face significant challenges in achieving complete decarbonization. CCUS can bridge this gap by capturing emissions from these sectors, ensuring a more comprehensive approach to reducing greenhouse gases [9–11].

Responsibly Meeting Energy Needs: Satisfying the world's ever-growing energy demands is a complex issue. While the shift toward renewable energy is underway, fossil fuels remain a significant part of our energy mix. CCUS offers a responsible way to meet energy needs while mitigating climate change. It captures CO_2 emissions from power generation and industrial processes, effectively reducing the carbon footprint of these energy sources and providing a crucial transition period toward cleaner alternatives.

Providing Negative Emissions: Perhaps one of the most compelling aspects of CCUS is its capacity to provide "negative emissions." Negative emissions technologies are those that actively remove CO_2 from the atmosphere.

CCUS technologies do this by capturing CO_2 emissions and storing them underground. This negative emission capacity is particularly valuable in sectors where achieving zero emissions is economically or technically infeasible, such as aviation or heavy industry. CCUS can play a pivotal role in offsetting emissions from these sectors, helping us balance the global carbon budget [12].

Reducing Emissions on Multiple Fronts: CCUS technologies are versatile, playing multiple strategic roles in the transition to a net-zero carbon future. Firstly, they significantly reduce emissions from power and industrial sectors by capturing CO_2 emissions at their source. Secondly, they enable the production of low-carbon hydrogen, which can replace fossil fuels in various applications, further reducing emissions. Thirdly, CCUS can directly remove CO_2 from the atmosphere through processes like direct air capture. These multifaceted roles make CCUS a cornerstone technology in the fight against climate change [13–15].

Meeting Climate Change Mitigation Goals: Achieving our climate change mitigation goals, such as those outlined in the Paris Agreement, requires a concerted effort across the globe. CCUS is an indispensable component of this effort. It provides a means of reducing emissions across sectors that are challenging to decarbonize entirely. Without CCUS, the path to carbon neutrality would be far more difficult and costly, making it an essential tool for achieving our climate objectives [16].

In conclusion, carbon capture, utilization, and storage (CCUS) represent a powerful and multifunctional technology with the potential to reshape our approach to mitigating climate change. By capturing CO_2 emissions from industrial processes and energy production, providing negative emissions, and enabling the transition to low-carbon energy sources, CCUS addresses multiple dimensions of the climate crisis. However, it is essential to recognize that CCUS should complement, not replace, other critical strategies like renewable energy adoption and energy efficiency improvements. As we work toward a sustainable and carbon-neutral future, CCUS stands as a crucial pillar in our collective efforts to combat climate change and protect the planet for future generations.

Leakage prevention: The primary concern in a carbon capture and storage (CCS) project is the potential release of CO_2 from the storage sites [17]. According to Benson et al. [18], the risk of such leakages is significantly elevated during the initial injection phase into a reservoir or field. This heightened risk primarily arises from the geological complexities involved and the limited data available to comprehensively understand the consequences of introducing CO_2 into a geological site. However, the knowledge and experience gained from the initial injection can contribute to safer subsequent injections into the same reservoir or field over time. Consequently, it is imperative to employ a robust risk assessment approach, particularly during the early stages of carbon storage. This approach ensures the safety and security of the storage site through proper site selection, characterization, and informed decision-making analysis [18].

Several measures should be taken to prevent carbon capture and storage (CCS) leakage. These includes risk assessment, a crucial step in preventing CCS leakage. In this regard, potential risks associated with CCS projects are identified, analyzed, and evaluated to understand the likelihood and consequences of various hazards and develop strategies to mitigate them [1].

Continuous monitoring of CCS sites to detect any potential leakage using various techniques such as seismic surveys, pressure monitoring, and gas sampling is pertinent [4].

The selection of an appropriate site is critical to prevent CCS leakage and should be done based on geological and hydrological characteristics that minimize the risk of leakage [3].

Proper well designed and construction are essential to prevent CCS leakage. Wells should be designed and constructed to withstand the high pressure and corrosive nature of CO_2 [4]. Injection and storage techniques play a crucial role in preventing CCS leakage. The right injection and storage techniques should be used to minimize the risk of leakage and ensure the safe and effective storage of CO_2 [2].

It is important to note that these measures should be implemented in conjunction with ongoing risk assessment and monitoring to ensure the safe and effective implementation of CCS technologies.

7.6 Consequences of Carbon Capture and Storage Leakage

Some of the potential consequences of carbon capture and storage (CCS) leakage include:

Environmental Impacts: CCS leakage can lead to environmental consequences. If stored carbon dioxide (CO_2) leaks into the atmosphere, it can contribute to greenhouse gas emissions and exacerbate climate change [5]. Additionally, if CO_2 leaks into groundwater, it can contaminate drinking water sources and harm ecosystems [5].

These leakages can pose health and safety risks to workers and nearby communities. If there is a sudden release of CO_2, it can displace oxygen and create hazardous conditions [1]. This can lead to asphyxiation and other health hazards.

It is important to note that the consequences of the leakage can vary depending on the scale and severity of the leakage, as well as the specific circumstances of each project. Implementing robust risk assessment, monitoring, and mitigation measures can help minimize the likelihood and consequences of CCS leakage.

7.7 Case Studies

Real-world examples of carbon capture, utilization, and storage (CCUS) projects with a focus on Health, Safety, and Environmental (HSE) practices include:

Gorgon CCS project: The Gorgon CCS project in Australia is one of the world's largest commercial-scale CCUS projects. The project captures CO_2 emissions from the Gorgon natural gas facility and stores them in a deep saline formation beneath Barrow Island. The project has implemented robust HSE practices, including a comprehensive risk management plan, regular safety training for personnel, and reliable detection systems [18].

Quest CCUS project: The Quest CCUS project in Canada is another large-scale CCUS project that captures CO_2 emissions from an oil sands upgrader and stores them in a deep saline formation. The project has implemented a risk-based approach to HSE, identifying and assessing potential risks and developing safety technologies to mitigate those risks [19].

Illinois Industrial CCS project: The Illinois Industrial CCS project is a large-scale CCUS project that captures CO_2 emissions from an ethanol plant and stores them in a deep saline formation. The project has developed a project risk assessment and monitoring report that includes guidance on hazard risk management specific to CO_2 safety and environmental major accidents for CCUS projects [20].

CCUS projects in Asia-Pacific: According to McKinsey, there is vast potential for CCUS in Asia-Pacific, and there has been tangible progress in Australia and China in this area. However, more needs to be done to establish a detailed government policy on national climate goals, fit-for-purpose operations, and HSE regulations. Governments could start supporting CCUS pilot studies with project-specific approvals and clear objectives around demonstrating technical viability; awarding provisional storage licenses; leveraging current petroleum regulations; and temporarily broadening the scope of existing government agencies [21].

CCUS projects in the global economy: A survey of technical assessment of CCUS projects in the global economy found that challenges of the CCUS system must be overcome to raise many low TRL. The survey identified the need for robust HSE practices, including hazard risk management specific to CO_2 safety and environmental major accidents for CCUS projects [22].

Petra Nova Coal Plant: The Petra Nova coal plant in Texas was once considered a poster child for CCS technology. However, the plant consistently underperformed, and it finally closed for good in 2020 [4].

San Juan Generating Station: The San Juan Generating Station in New Mexico, touted as the largest capture project in the world, may already be headed for a similar fate [4].

Aliso Canyon Gas Leak: The California Aliso Canyon gas leak in 2015 was the worst man-made greenhouse gas disaster in US history, when 97,000 metric tons of methane leaked into the atmosphere. While the leak at Aliso Canyon was a methane, not carbon dioxide leak, depleted oil and gas reservoirs are commonly used to store captured carbon dioxide [3].

It is important to note that these incidents are relatively rare, and CCS technologies have been developed to minimize the risk of leakage. Ongoing research and development efforts aim to improve the safety and effectiveness of CCS technologies and minimize the likelihood of leakage incidents.

Overall, these examples demonstrate the importance of implementing robust HSE practices in CCUS projects to ensure the safe and successful operation of CCUS facilities.

7.8 Causes of Carbon Capture and Storage Leakage Incidents

Poor Performance: CCS technologies have a long history of failure and underperformance [1]. Poor performance can lead to leakage incidents, as the technology may not be able to capture and store carbon dioxide (CO_2) effectively.

Storage Location: Storage locations can leak CO_2, as they are often sited near fossil fuel reservoirs [6]. Oil and gas wellbores provide a pathway for CO_2 to escape to the surface, which can contaminate groundwater and soil.

Sudden Catastrophic Leak: A sudden catastrophic leak can be caused by a well blowout or pipeline rupture [4]. This kind of leak can release large amounts of CO_2 into the atmosphere and have local effects such as ground and water displacement, groundwater contamination, and biological interactions.

It is important to note that these causes can vary depending on the specific circumstances of each project. Implementing a robust risk assessment, monitoring, and mitigation measures can help minimize the likelihood and consequences of CCS leakage incidents.

7.9 Best Practices

Measurement, Monitoring, and Verification (MMV) Processes: Assuring the safe storage of CO_2 requires robust measurement, monitoring, and verification (MMV) processes to be in place that would enable any CO_2 leaks from a storage site to be detected and quantified [5].

Continuous Monitoring: Continuous monitoring of CCS sites is essential to detect any potential leakage. Monitoring can be done using various techniques such as seismic surveys, pressure monitoring, and gas sampling [1, 2, 5].

Robust Methodology and Cost-Effective Tools: The Strategies for Environmental Monitoring of Marine Carbon Capture and Storage (STEMM-CCS) project aims to provide a robust methodology and cost-effective tools to identify, detect, and quantify CO_2 leakage in the unlikely event it occurs from a sub-seafloor CCS reservoir.

Integration of Qualitative and Quantitative Risk Assessment Methods: Integrating qualitative and quantitative risk assessment methods can help improve the accuracy of monitoring and verification processes [3].

It is important to note that these best practices should be implemented in conjunction with ongoing risk assessment and mitigation measures to ensure the safe and effective implementation of CCS technologies.

7.10 Key Components of a Monitoring Plan

Establishing Monitoring Objectives: The monitoring plan should clearly define the objectives and goals of the monitoring activities. This includes identifying the specific parameters to be monitored, such as CO_2 concentrations, pressure, and temperature [4].

Baseline Monitoring: Baseline monitoring involves collecting data on the preoperational conditions of the storage site. This establishes a reference point against which future monitoring data can be compared to detect any changes or potential leakage [4].

Continuous Monitoring: Continuous monitoring is essential to detect any potential leakage and ensure the ongoing integrity of the storage site. This can involve various techniques such as seismic surveys, pressure monitoring, and gas sampling [1, 3].

Data Collection and Analysis: The monitoring plan should outline the methods and frequency of data collection, as well as the analysis techniques to be used. This includes establishing thresholds or triggers for action based on the monitoring data [4].

Reporting and Communication: The monitoring plan should include provisions for regular reporting and communication of monitoring results. This ensures that relevant stakeholders are informed about the status of the CCS project and any potential issues that may arise [4].

Adaptive Management: The monitoring plan should incorporate adaptive management principles, allowing for adjustments and improvements based on the monitoring results. This ensures that the monitoring plan remains effective and responsive to changing conditions [4].

It is important to note that the specific components of a monitoring plan may vary depending on the specific CCS project and regulatory requirements. The plan should be tailored to the unique characteristics and risks associated with each storage site.

7.11 HSE Practices in CCUS Projects

The implementation of robust health, safety, and environmental (HSE) practices is paramount to ensure the secure and efficient operation of CCUS facilities. Here, we outline key best practices for maintaining rigorous HSE standards in the context of CCUS:

Adoption of a Risk-Based Approach: A risk-based approach involves systematically assessing the various risks associated with CCUS operations. This includes evaluating the geographic locations, specific activities, and services that may pose higher risks. By doing so, companies can allocate their resources more efficiently to address these higher-risk areas.

This approach facilitates the development of targeted HSE standards. These standards are typically more concise, making them easier for personnel to understand and follow. Moreover, they can be continuously monitored for their effectiveness and adherence to compliance requirements.

Ultimately, a risk-based approach helps companies prioritize their efforts to minimize potential hazards and reduce the likelihood of incidents, promoting safer CCUS operations [23–26].

Ensuring a Dependable Detection Systems: Reliable detection systems are essential for worker safety within CCUS facilities. These systems are responsible for identifying leaks, emissions, or abnormal conditions promptly. Adequate sensor technology and monitoring equipment should be in place to detect the presence of CO_2 or other hazardous substances. Early detection is crucial to taking swift corrective actions and preventing accidents. Additionally, regular maintenance and calibration of these detection systems are necessary to ensure their continued reliability [27, 28].

Sharing Global Best Practices and Experiences: The CCUS industry can greatly benefit from the exchange of knowledge and experiences on a global scale. Sharing best practices, case studies, and lessons learned from different regions can enhance the overall effectiveness and safety of CCUS projects. Collaboration among governments, industry players, and research institutions is crucial to facilitate this knowledge sharing. It helps in overcoming challenges and streamlining the implementation of CCUS technology [19].

Identifying and Mitigating Health and Safety Risks: The comprehensive identification and mitigation of health and safety risks are fundamental to safeguarding the operation of CCUS facilities. Risks can arise at various stages of the CCUS process, from carbon capture to transportation and underground storage. Hazard assessments and risk mitigation strategies must be developed and regularly updated to address evolving threats. Adequate training and preparedness of personnel are also essential components of risk mitigation. This includes emergency response plans and drills [29–31].

Government Commitment to Safe CCUS Advancement: Governments play a pivotal role in promoting the safe advancement of CCUS technology. They can establish regulatory frameworks and standards that prioritize safety. Committing to the safe development and deployment of CCUS technology at the regional or national level demonstrates a government's dedication to addressing climate change while ensuring public safety.

Implementing Best Management Practices (BMPs): Best Management Practices (BMPs) provide a set of guidelines and standards specifically tailored to offshore transportation and sub-seabed geologic storage of carbon dioxide. BMPs cover various aspects, including site selection, well design, monitoring, and risk management. By following these practices, operators can minimize environmental risks and ensure the secure containment of CO_2 [32].

In summary, these best practices collectively form a comprehensive approach to ensuring the safety, effectiveness, and responsible implementation of CCUS technology. By embracing these principles, the CCUS industry can contribute significantly to reducing carbon emissions and combatting climate change while prioritizing the well-being of workers and the environment.

7.12 Identifying and Assessing Risks in CCUS Projects

Recognizing and appraising risks constitutes a pivotal phase in the progression of carbon capture, utilization, and storage (CCUS) initiatives. Here are several methods for recognizing and assessing risks within CCUS projects:

Establishing a Comprehensive Risk Assessment Program: The establishment of a dedicated risk assessment program is imperative for pinpointing and appraising risks within CCUS projects. This program should be tailored specifically to the intricacies of CCUS but can be adaptable to other project types as well. It should encompass a thorough comprehension of the sequence of events that may culminate in significant setbacks within a CCUS endeavour [33].

Identifying Project Risks: The initial identification of project risks typically transpires at the inception of the project, often during brainstorming sessions. These sessions can be dedicated to pinpointing the primary risks associated with the undertaking, encompassing technical, environmental, social, and financial aspects [34].

Formalizing an Integrated Decision-Making Model: Existing theoretical research tend to overlook the importance of comprehensive risk assessments for CCUS projects from a sustainability perspective. Consequently, it is imperative to formalize an integrated decision-making model to assess the risks associated with CCUS projects from a sustainability standpoint [35].

Analyzing and Evaluating Potential Risks: Risk assessment entails an overarching process that involves identifying, analyzing, and evaluating potential risks. In the context of CCUS projects, this encompasses a meticulous analysis and evaluation of health, safety, and environmental risks [35].

Identifying Key Risks Affecting Public Confidence: The identification of pivotal risks that impact public confidence in CCUS initiatives and government decision-making processes is a requisite step in formulating a risk management strategy [35].

In summary, the identification and assessment of risks within CCUS projects are of paramount importance to ensure the secure and prosperous operation of CCUS facilities. Employing a risk assessment program, recognizing project-specific risks, formalizing an integrated sustainability-driven decision-making model, scrutinizing, and appraising potential risks, and identifying key factors that influence public trust are all effective techniques for identifying and evaluating risks inherent to CCUS projects.

7.13 Public Concerns in CCS

The public has concerns toward the developments towards CCS and below are some primary apprehensions held by the public concerning CCS:

Limited Awareness and Understanding: Public awareness of CCS remains relatively low, and the general knowledge about CCS technology prior to in-depth discussions and diverse viewpoints does not consistently indicate approval or disapproval [36, 37].

Safety Issues: Concerns revolve around the safety of CO_2 transportation and sequestration, with worries about potential CO_2 injection-triggered earthquakes, which have been documented at some injection sites. An incident involving a ruptured CO_2 pipeline in a predominantly Black community in Mississippi resulted in residents seeking medical treatment and harmed local flora and fauna [38, 39].

Financial Considerations: CCS is a costly technology, illustrated by the US$6.9 billion spent by the U.S. Department of Energy between 2005 and 2012 in an attempt to demonstrate its feasibility for coal. However, this substantial investment yielded limited tangible results, as less than 4% of the planned CCS capacity was deployed from 2014 to 2016 [37].

Fossil Fuel Dependence: A prevailing concern among many is that CCS could inadvertently foster increased reliance on fossil fuels.

CCS Viability: Persistent doubts linger about the long-term effectiveness and sustainability of CCS in reducing carbon emissions.

Overall, it becomes evident that the public's principal reservations regarding CCS relate to issues of safety, financial feasibility, and the technology's efficacy in mitigating carbon emissions.

7.14 Conclusion

This chapter discusses the identification and assessment of risks within CCUS projects are of paramount importance to ensure the secure and prosperous operation of CCUS facilities. Employing a risk assessment program, recognizing project-specific risks, formalizing an integrated sustainability-driven decision-making model, scrutinizing and appraising potential risks, and identifying key factors that influence public trust are all effective techniques for identifying and evaluating risks inherent to CCUS projects.

References

[1] Oraee-Mirzamani, B., Cockerill, T. and Makuch, Z. 2013. Risk assessment and management associated with CCS. Energy Procedia 37: 4757–4764. doi: 10.1016/j.egypro.2013.06.385.
[2] Christopher F. Brown, Greg Lackey, Brandon Schwartz, Marcella Dean, Robert Dilmore, Hein Blanke et al. 2022. 16th International Conference on Greenhouse Gas Control Technologies GHGT-16 Extended abstract to : Integrating Qualitative and Quantitative Risk Assessment Methods for Carbon Storage: A Case Study for the Quest Carbon Capture and Storage Facility. no. October, 2022.
[3] Date, P. 2007. Lawrence Berkeley National Laboratory.
[4] Wilday, J., Paltrinieri, N., Farret, R., Hebrard, J. and Breedveld, L. 2011. Carbon Capture and Storage: A Case Study of Emerging Risk Issues in The. 156: 339–346.
[5] Whitfield, A. 2011. An environmental risk assessment for carbon capture and storage. Inst. Chem. Eng. Symp. Ser. 156: 1–7.
[6] Jiang, K., Ashworth, P., Zhang, S., Liang, X., Sun, Y. and Angus, D. 2020. China's carbon capture, utilization and storage (CCUS) policy: A critical review. Renew. Sustain. Energy Rev. 119(April 2019): 109601. doi: 10.1016/j.rser.2019.109601.
[7] Choi, J.-Y. and Byeon, S.-H. 2020. HAZOP methodology based on the health, safety, and environment engineering. International Journal of Environmental Research and Public Health 17(9). doi: 10.3390/ijerph17093236.
[8] Mohammadfam, I., Mahmoudi, S. and Kianfar, A. 2012. Development of the health, safety and environment excellence instrument: a HSE-MS performance measurement tool. Procedia Eng. 45: 194–198. doi: https://doi.org/10.1016/j.proeng.2012.08.142.
[9] Mac Dowell, N., Fennell, P. S., Shah, N. and Maitland, G. C. 2017. Mitigating climate change. Nat. Publ. Gr., no. April, 2017, doi: 10.1038/nclimate3231.
[10] Li, Q., Liu, G., Leamon, G., Liu, L., Cai, B. and Chen, Z. 2017. A national survey of public awareness of the environmental impact and management of CCUS technology in China. 114(November 2016): 7237–7244. doi: 10.1016/j.egypro.2017.03.1854.
[11] Jiang, K., Ashworth, P., Zhang, S., Liang, X., Sun, Y. and Angus, D. 2020. China's carbon capture, utilization and storage (CCUS) policy: A critical review. Renew. Sustain. Energy Rev. 119: 109601. doi: https://doi.org/10.1016/j.rser.2019.109601.
[12] Hashemi, S. E. 2021. A fuzzy multi-objective optimization model for a sustainable reverse logistics network design of municipal waste-collecting considering the reduction of emissions. J. Clean. Prod. 318(August): 128577. doi: 10.1016/j.jclepro.2021.128577.
[13] Jin, Y., Andersson, H. and Zhang, S. 2016. Air pollution control policies in China: A retrospective and prospects. doi: 10.3390/ijerph13121219.
[14] Yi, Q., Li, C., Tang, Y. and Member, S. 2012. A new operational framework to job shop scheduling for reducing carbon emissions.
[15] Valderrama, C. V., Santibanez-González, E., Pimentel, B., Candia-Véjar, A. and Canales-Bustos, L. 2020. Designing an environmental supply chain network in the mining industry to reduce carbon emissions. J. Clean. Prod. 254: 119688 doi: https://.doi.org/10.1016/j.jclepro.2019.119688.
[16] Placeholder, C. 2016. Carbon Capture, Utilization, and Storage: Climate Change, Economic Competitiveness, and Energy Security Carbon Capture, Utilization, and Storage: Climate Change, Economic Competitiveness, and Energy Security. no. August, 2016.
[17] Deel, D., Mahajan, K., Mahoney, C. R., McIlvried, H. G. and Srivastava, R. D. 2007. Risk assessment and management for long-term storage of CO_2 in geologic formations-United States Department of Energy R&D. Syst. Cybern. Informatics 5(1): 79–84.
[18] Benson, S. M. 2007. Confidence building in CCS: the role of industrial analogues.

[19] Iea, A. and Handbook, C. December 2022. CO_2 Storage Resources and their Development.
[20] Whittaker, S. and Carman, C. 2022. CarbonSAFE Illinois-Macon County Final Report. Univ. of Illinois at Urbana-Champaign, IL (United States).
[21] Rassool, D. 2021. Unlocking private finance to support CCS investments. Glob. CCS Inst.
[22] Dziejarski, B., Krzyżyńska, R. and Andersson, K. 2023. Current status of carbon capture, utilization, and storage technologies in the global economy: A survey of technical assessment. Fuel 342: 127776. doi: https://doi.org/10.1016/j.fuel.2023.127776.
[23] Mohammadfam, I., Nikoomaram, H. and Faridan, M. 2013. Evaluation of health, safety and environment (HSE) culture. Int. J. Occup. Hyg. 5(1): 1–5.
[24] Høivik, D., Moen, B. E., Mearns, K. and Haukelid, K. 2009. An explorative study of health, safety and environment culture in a Norwegian petroleum company. Saf. Sci. 47(7): 992–1001.
[25] Li, W., Liang, W., Zhang, L. and Tang, Q. 2015. Performance assessment system of health, safety and environment based on experts' weights and fuzzy comprehensive evaluation. J. Loss Prev. Process Ind. 35: 95–103.
[26] Poursadeqiyan, M. and Arefi, M. F. 2020. Health, safety, and environmental status of Iranian school: A systematic review. J. Educ. Health Promot. 9.
[27] Tang, H., Zhang, S. and Chen, W. 2021. Assessing representative CCUS layouts for China's power sector toward carbon neutrality. Environ. Sci. Technol. 55(16): 11225–11235.
[28] Xiao, K., Yu, B., Cheng, L., Li, F. and Fang, D. 2022. The effects of CCUS combined with renewable energy penetration under the carbon peak by an SD-CGE model: Evidence from China. Appl. Energy 321: 119396.
[29] Shavalieva, G., Kazepidis, P., Papadopoulos, A. I., Seferlis, P. and Papadokonstantakis, S. 2021. Environmental, health and safety assessment of post-combustion CO_2 capture processes with phase-change solvents. Sustain. Prod. Consum. 25: 60–76.
[30] Badr, S., Frutiger, J., Hungerbuehler, K. and Papadokonstantakis, S. 2017. A framework for the environmental, health and safety hazard assessment for amine-based post combustion CO_2 capture. Int. J. Greenh. Gas Control 56: 202–220.
[31] Fogarty, J. and McCally, M. 2010. Health and safety risks of carbon capture and storage. JAMA 303(1): 67–68.
[32] Boem, O. C. S. S. 2018. Best Management Practices for Offshore Transportation and Sub-Seabed Geologic Storage of Carbon Dioxide Best Management Practices for Offshore Transportation and Sub-Seabed Geologic Storage of Carbon Dioxide.
[33] Oraee-Mirzamani, B., Cockerill, T. and Makuch, Z. 2013. Risk assessment and management associated with CCS. Energy Procedia 37: 4757–4764. doi: https://doi.org/10.1016/j.egypro.2013.06.385.
[34] Report, N. 2012. "PROJECT PIONEER," Preliminary construction and execution strategy a summary produced for global CCS institutre 2012.
[35] Liu, B., Liu, S., Xue, B., Lu, S. and Yang, Y. 2021. Formalizing an integrated decision-making model for the risk assessment of carbon capture, utilization, and storage projects: From a sustainability perspective. Appl. Energy 303: 117624. doi: https://doi.org/10.1016/j.apenergy.2021.117624.
[36] Tcvetkov, P., Cherepovitsyn, A. and Fedoseev, S. 2019. Public perception of carbon capture and storage: A state-of-the-art overview. Heliyon 5(12): e02845. doi: 10.1016/j.heliyon.2019.e02845.
[37] Whitmarsh, L., Xenias, D. and Jones, C. R. 2019. Framing effects on public support for carbon capture and storage. Palgrave Commun. 5(1). doi: 10.1057/s41599-019-0217-x.

[38] Terwel, B. W., Harinck, F., Ellemers, N. and Daamen, D. D. L. 2011. Going beyond the properties of CO_2 capture and storage (CCS) technology: How trust in stakeholders affects public acceptance of CCS. Int. J. Greenh. Gas Control 5(2): 181–188.
[39] Schumann, D. 2015. Public acceptance. Carbon Capture, Storage Use Tech. Econ. Environ. Soc. Perspect., pp. 221–251.

CHAPTER 8

Methods for Modelling and Simulating CO_2 Capture Utilizing Hydrate Technology

Vishal Srivastava[1,*] and *Bhajan Lal*[2]

8.1 Introduction

CO_2 capture technologies encompass a range of strategies designed to reduce the release of carbon dioxide into the atmosphere, contributing to the global effort to combat climate change. These technologies are crucial for mitigating the impact of industrial processes and energy production on the environment. Among the various methods available, hydrate-based approaches have emerged as a particularly promising avenue. One compelling factor contributing to the significant potential of hydrate-based CO_2 capture lies in its ability to effectively capture and securely store CO_2. This dual capability not only aids in the capture of CO_2 emissions but also establishes a reliable storage solution. This addresses a significant challenge in CO_2 capture, as finding suitable and secure storage sites for captured CO_2 is a complex task.

However, one of the foremost challenges associated with hydrate-based CO_2 capture lies in the realm of modelling and simulation. The complexity of hydrate-based CO_2 capture processes adds layers of intricacy to CO_2 hydrate formation modelling efforts. A precise depiction of the intricate interaction among variables such as temperature, pressure,

[1] New York Life's Center for Data Science and Artificial Intelligence (CDSAi), New York City Metropolitan Area.

[2] Chemical Engineering Department, Center for Carbon Capture, Storage and Utilization (CCUS), Institute of Sustainable Energy, Universiti Teknologi PETRONAS, Bandar Seri Iskandar, Perak, Malaysia.

* Corresponding author: vshfrm@gmail.com

gas composition, and molecular interactions requires a comprehensive grasp of modelling techniques and the fundamental mechanisms at play. Moreover, since hydrate formation involves multiple phases, ensuring that these phases are adequately captured in simulations is no trivial task. The challenge becomes especially pronounced when aiming to scale up from lab-scale experiments to industrial applications. The transition introduces additional complexities tied to fluid dynamics, mass transfer heat transfer, and reactor kinetics.

This chapter primarily encapsulates the intricacies associated with the modelling of hydrate formation to provide actionable insights for large-scale implementation. Therefore, the objective here is to create computational models that depict the intricate processes and fundamental mechanisms underlying CO_2 hydrate formation. This endeavour aims to foster a comprehensive understanding of quantitative modelling pertaining to the formation of CO_2 hydrates, paving the way for advancements in carbon capture. Ultimately, the insights from this chapter can be instrumental in predicting the behaviour of CO_2 hydrate-based systems under various conditions and optimizing their performance.

Nevertheless, it is crucial to acknowledge that hydrate modelling encompasses both time-independent and time-dependent facets, and the distinction relies on whether one adopts a thermodynamic or kinetic viewpoint. In the realm of thermodynamics, time does not play a pivotal role as the processes ultimately reach a stable state. Conversely, kinetic modelling delves into the dynamic aspects of hydrate formation rates and the changing characteristics of hydrate formation and dissociation, influenced by factors like nucleation and growth. The following section provides brief overviews of both perspectives.

8.2 Two Primary Dimensions to Hydrate Modelling

Hydrate modelling encompasses two fundamental aspects: thermodynamic modelling and kinetic modelling. A concise description of each aspect is discussed below.

8.2.1 Thermodynamic Modelling

Hydrate formation and dissociation are driven by process drivers including pressure and temperature which lead to the thermodynamic equilibrium, which means that the process strives to reach a stable state over time. However, the time it takes to reach this equilibrium depends on the system's conditions. The system must exist within the hydrate formation region, as mandated by the system thermodynamics. In this chapter, we delve deeper into the subject of kinetic modelling for hydrate formation, specifically focusing on hydrate nucleation and growth processes.

8.2.2 Hydrate Nucleation and Growth Modelling

Hydrate formation initiates with nucleation, characterized by the emergence of the first hydrate crystals at specific locations within a mixture of gas and water. As discussed by [1] hydrate nucleation represents a microscopic phenomenon wherein tiny hydrate crystals, often referred to as nuclei, undergo growth and dispersion as they endeavour to reach a critical size necessary for further expansion. Further, as elucidated by [1–2] in the presence of both hydrate former and water in the hydrate stability zone, the immediate formation of hydrate does not occur. Rather, the process of hydrate formation unfolds over time and is governed by the heat or mass transfer of constituent elements to the growing crystal surface, as well as the reactions kinetics. The process of hydrate formation and growth is therefore time dependent. The rate of hydrate formation is influenced by various factors, including temperature, pressure, and the gas and water purity. In this chapter, we delve into the quantitative aspects of each of these phases, specifically within the context of carbon dioxide hydrate formation.

8.3 Modelling Facets to Hydrate Nucleation

In a typical experiment designed for hydrate formation, a hydrate-forming gas is introduced into water under specified room temperature and pressure conditions. Subsequently, the system is gradually cooled to a temperature lower than the equilibrium hydrate formation temperature, denoted as T_{eq}. Researchers often employ subcooling, defined as the difference between T_{eq} and the current system temperature (T_{system}), to determine the driving force responsible for hydrate nucleation and growth.

The interval from when the system temperature falls below T_{eq} to when hydrate formation becomes evident is referred to as the induction period. During this induction period, the system is thermodynamically situated within the hydrate formation region, but nucleation of hydrates is

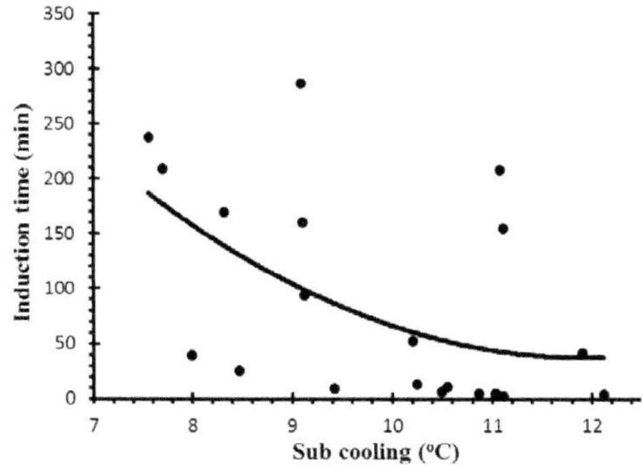

Figure 1. The relationship between induction time and subcooling [3] for mixed hydrate containing C_1/C_3. The solid line on the right side of the graph is provided as a visual aid.

not observed due to the significant metastability exhibited by the system. However, data suggests that the induction time generally decreases with increasing subcooling, as expected. Furthermore, it is apparent that the onset of hydrate nucleation is not experimentally reproducible, as the nucleation process is stochastic in nature.

In the context of CO_2 hydrate formation, in a recent study by [4], using HP-ALTA, nucleation rates of CO_2 hydrate and methane – propane mixed hydrate were measured and compared, as illustrated below:

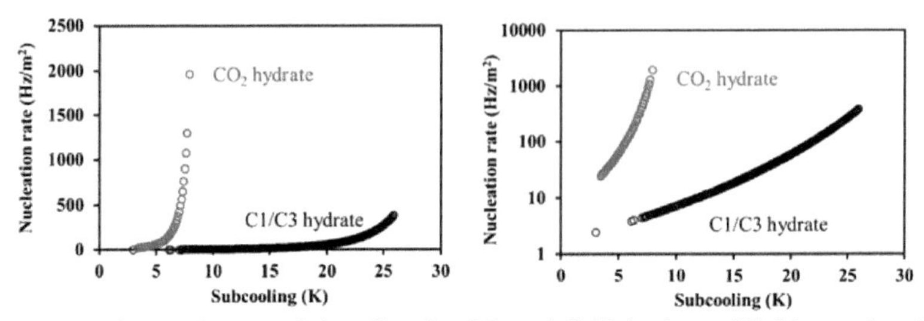

Figure 2. Nucleation Rate vs. Subcooling for CO_2 and C_1/C_3 hydrates [4]. Linear plots (left) and semi-logarithmic (right) plots are shown.

The result showed that nucleation rate for CO_2 hydrate was significantly higher than the methane–propane (C_1/C_3) mixed gas hydrate. Additionally, the data supported the hypothesis that nucleation rate increased as subcooling increased.

It was determined that the most suitable method for assessing the nucleation rate in heterogeneous nucleation involves measuring it as the rate per unit length (Hz/m^2) of the three-phase boundary. This choice was made based on the reasoning that the nucleation rate per unit length of this boundary is a distinctive characteristic of the specific solid material under investigation. Consequently, the observed nucleation rates per unit length displayed a non-linear increase, spanning from approximately 5 K to around 28 K in terms of the degree of subcooling. This development represents a significant step forward in the pursuit of accurately determining nucleation rates for carbon dioxide and mixed hydrates.

8.4 Driving Force for Homogeneous Nucleation

In their work, [5] first introduced the most general scenario for hydrate nucleation. They calculated the Gibbs free energy for hydrate nucleation under isothermal conditions, and their model suggests the following:

$$\Delta g = v_L\left(P_{eq} - P\right) + RT\sum x_i \ln\left(\frac{f_i^{eq}}{f_i}\right) + v_H\left(P - P_{eq}\right) \tag{1}$$

where Δg is the molar Gibb's free energy difference equivalent to the driving force for nucleation, v_L and v_H are molar volumes for liquid and hydrate respectively, P and P_{eq} are system pressure and hydrate equilibrium pressure respectively, x_i is the mole fraction of component i, and f_i and f_i^{eq} are fugacities of component i under experimental conditions and at equilibrium respectively.

[6] conducted an analysis to calculate the energy required for the formation of clusters of the new phase in both homogeneous and heterogeneous environments. They demonstrated that in practical situations, such as when the water is not sufficiently pure, hydrate nucleation tends to occur preferentially on solid substrates present in the solution. Additionally, they developed a formula for the rate of nucleation in single-component gas hydrates and investigated the impact of kinetic inhibitors on hydrate nucleation. According to their model, a suitable inhibitor should adsorb strongly to the nucleation site without introducing new nucleation sites. The single-component hydrate models were subsequently expanded to encompass multicomponent systems by [7–8]. According to their nucleation model, the Gibbs free energy difference or the driving force for nucleation is given by:

$$\Delta g = n_w \left[v_w \left(P - P_{eq} \right) - v_{hw} \left(P - P_{eq} \right) - k_B T \sum_j v_j \ln \left(\frac{1 - \sum_k \theta_{kj} \left(T, P, y \right)}{1 - \sum_k \theta_{kj} \left(T, P_{eq}, y \right)} \right) \right] \tag{2}$$

where, n_w is the number of water molecules in a unit cell, v_w is the molar volume of water molecules in the solution, v_{hw} is the molecular volume of water in the hydrate (hydrate unit cell volume divided by number of water molecules in the unit cell), P is the pressure, T is the temperature, k_B is the Boltzmann constant, v_j is the number of type j cavities per water molecule, y is the composition in the gas phase. θ_{kj} is the fractional filling of cavity j by a type k molecule expressed as:

$$\theta_{kj} = \frac{C_{kj} f_{gk}}{1 + \sum_i C_{ij} f_{gj}} \tag{3}$$

where C_{kj} is the Langmuir constant for species k in cavity j, f_{gk} is the fugacity of gas component k in the gas phase in equilibrium with the hydrate, and the summation is over all species except water.

This model demonstrates that apart from temperature, pressure, and gas-phase composition, the driving force for the nucleation of multicomponent gas hydrates is also influenced by the composition of the hydrate phase. Through the application of this model, [8] successfully anticipated nucleation rates and induction times for mixed systems. However, it is important to acknowledge that the models outlined in this section are built on the foundation of steady-state nucleation and, as a result, do not provide insights into the initial stages of nucleation. To explore the initial phases of nucleation, molecular simulations, which we will shortly delve into providing the most appropriate method.

8.5 Mechanistic Understanding of Nucleation using Molecular Simulations

In recent microsecond MD simulations of methane hydrate nucleation investigated the emergence of hydrate cages has been observed as guest molecules engage with the unstable ring structures within water, providing support for the labile cluster hypothesis. It was founded upon the labile cluster nucleation proposition, which suggests that labile ring formations are present in pure liquid water. When a hydrate-former dissolves, water molecules assemble into labile clusters around the guest molecule, ultimately culminating in the formation of hydrate structures [9].

Additionally, CO_2 deviates from the usual non-polar hydrocarbon hydrate formers by engaging in chemical interactions with water molecules. Consequently, it is plausible that the nucleation mechanism for CO_2 hydrates can be different among guest molecules and is influenced by their chemical properties. Indeed [11] explored this aspect and uncovered notable differences in the nucleation processes of hydrates derived from hydrophobic and hydrophilic hydrate formers.

Additionally, [12] introduced a novel concept that combines elements of the labile cluster and local structuring theories, known as the "blob hypothesis." According to the blob hypothesis, the initial step involves the reversible formation of "blobs," which are enduring assemblies of guest molecules separated by water molecules. Within these blobs, clathrate cages continuously form and dissolve until a cluster of cages attains a critical size. Once this critical size is achieved, further growth occurs through the space-filling mechanism, involving the sharing of faces among clathrate cages.

The key distinction between a "blob" and an amorphous clathrate lies in the fact that, within a blob, water molecules are not confined within clathrate cages, whereas in an amorphous crystal, there are hydrogen-bonded polyhedral cages that rigidly define the structure.

The proper selection of order parameters is crucial for rare event sampling simulation techniques. Recently, Barnes et al. conducted

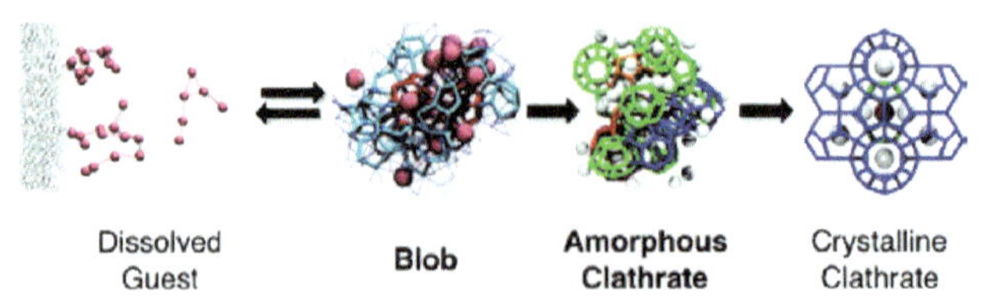

Figure 3. The blob hypothesis pertaining to hydrate nucleation [11].

equilibrium path sampling calculations and identified the Mutually Coordinated Guest (MCG) order parameter as the appropriate reaction coordinate through the pB histogram test. By testing the MCG order parameter, they demonstrated the structural organization of both guest (methane) and host (water) [14]. They also identified a critical nucleus size of MCG-1 = 16 at 255 K and 500 bar based on their simulations. Their proposed nucleation mechanism, as depicted in Figure 6, involves green spheres representing guest molecules that are mutually coordinated with water clusters (red). These structures undergo stepwise, nondeterministic growth and rearrangement. It is worth noting that the connectivities illustrated in Figure 4 below are for explanatory purposes, and in reality, clusters of specific sizes can possess various connective patterns.

Simulations conducted by Barnes et al. [14] and Bi et al. [13] indicate that the free energy profile associated with hydrate nucleation can be adequately explained by classical nucleation theory. This finding is somewhat unexpected, considering the non-classical molecular nucleation pathway typically associated with hydrates.

However, it is important to note that accurate data regarding the crystal growth rate after hydrate nucleation are quite limited. Many of the crucial factors that impact nucleation, such as deviations from equilibrium conditions, surface area, agitation, the history of water, and the composition of the gas, continue to be significant as the hydrate growth phase ensues.

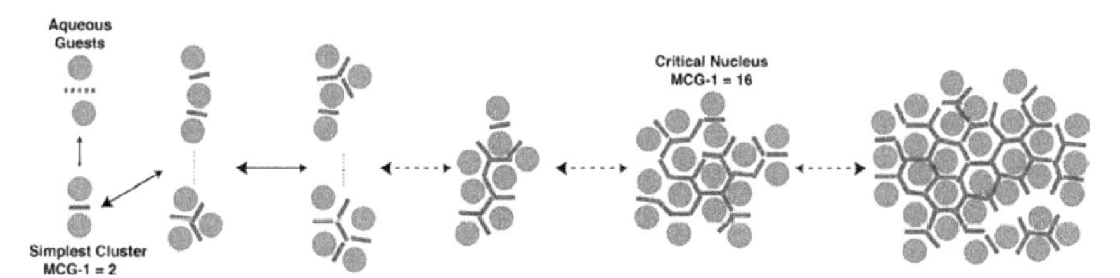

Figure 4. Hydrate nucleation mechanism based on the guest order parameter that is mutually coordinated as depicted by Barnes et al. [14].

8.6 Hydrate Growth

Recognizing the inherently unpredictable nature of hydrate crystal nucleation, the ability to predict the rate at which hydrate formation takes place offers a practical solution when engaging in hydrate formation modelling. Quantifying the rate of hydrate growth not only deepens our comprehension of hydrate formation but also streamlines the quantification and modelling procedures, contributing to the development of a more dependable framework for forecasting CO_2 hydrate formation.

From an industrial perspective, the rate at which hydrate formation occurs is critical, as it directly influences various aspects of the process. The figure below illustrates various stages of hydrate formation, highlighting

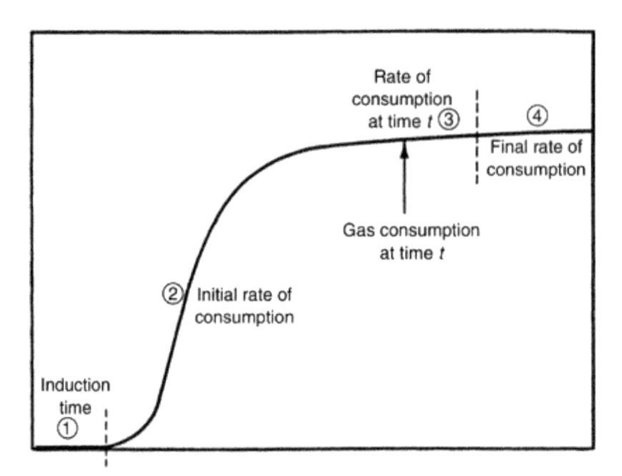

Figure 5. Symbolic representation to depict a typical profile for the consumption of hydrate former (gas) with the passage of time, taken with permission from Sloan and Koh [2].

the significance of understanding hydrate growth for practical modelling applications.

In the initial phase, commonly referred to as the "induction period," the temperature and pressure conditions fall within the range where hydrate formation is thermodynamically possible. However, despite this possibility, hydrate formation does not occur during this period due to a phenomenon known as metastability. Metastability is the ability of a nonequilibrium state to persist for an extended duration.

The subsequent phase, identified as the "growth period" within region 2, is characterized by a remarkably swift development of hydrates. During this growth phase, gas molecules become concentrated within the hydrate cages, where they are densely packed. As the process consumes water through hydrate formation, the rate at which gas is consumed gradually decreases over time, as depicted in the reduction of slope when we move from Points 3 to 4.

Nonetheless, as the growth process unfolds, the significance of both mass and heat transfer becomes increasingly prominent. In Figure 5 above, which illustrates the gas consumption resulting from the interaction between water and gas within a sealed system, the growth phase is marked as "2". During this phase, a substantial volume of gas becomes integrated into the hydrate structure. Several factors such as the solubility of CO_2 in water and the transport of CO_2 to the hydrate surface becomes of paramount significance and may even play a substantial role in governing the overall growth process. Additionally, the exothermic heat released during hydrate formation can also play a substantial role in governing the growth process. It is for these reasons that the concept of mass or heat transfer or hydrate formation kinetics become predominantly important for hydrate growth processes.

8.7 Overview of Hydrate Growth Models

Several research groups have explored three unique hydrate growth models, each contingent on three distinct types of resistances or rate limitations, as illustrated below:

A. Intrinsic growth kinetics.
B. Heat transfer limited.
C. Mass transfer limited.

This section presents a description of each type of the model with a brief critique. However, one should approach the use of any hydrate growth model with some caution, for the main reason that hydrate nucleation—the initiation of hydrate onset, occurring during the induction period—is a stochastic process, which includes significant uncertainty at a low driving force under isothermal conditions. Nevertheless, when we delve into the mechanistic understanding of hydrate growth, considering factors like intrinsic kinetics, heat transfer, and mass transfer resistances, it equips us with a practical toolbox that allows for reasonably accurate predictions of hydrate growth rates in practical applications. Therefore, a more in-depth exploration of each of the three models is presented here.

Intrinsic Growth Kinetic Model

Hydrate formation processes may face inherent limitations stemming from the kinetics of the reaction itself, and it is important to delve into the specifics of these limitations. Intrinsic kinetic limitations refer to the inherent rate at which hydrates form under given conditions. These limitations are fundamentally linked to the inherent kinetics of the hydrate formation reaction. When hydrate formation is kinetic limited, it means that the reaction cannot proceed any faster due to the nature of the molecules involved and the activation energy required for the formation process. Being the pioneering work initially presented by Englezos et al., they observed that the hydrate formation rate constant K* exhibited Arrhenius temperature dependence [2].

During the same timeframe, Kim et al. examined the kinetics of methane hydrate decomposition by employing a semi-batch stirred-tank reactor. Their findings revealed that the decomposition rate was directly proportional to the particle surface area and the disparity between the methane fugacity at equilibrium pressure and the decomposition pressure [15]. They pointed out that the rate constant K^* showed Arrhenius temperature dependence and introduced a kinetic model for hydrate decomposition. Subsequently, in the study by Moridis et al., a kinetic framework for hydrate formation was adopted, as shown below [16].

$$K^{*} = -K_{O} Exp\left(\frac{\Delta E}{RT}\right) \tag{4}$$

Their findings further indicated that the decomposition rate was directly proportional to the particle's surface area and the disparity between the fugacity of methane at equilibrium pressure and the decomposition pressure.

$$\frac{dn_i}{dt} = K^{*} A_p (f_{bi} - f_{eqi}) \tag{5}$$

Where: $\frac{dn_i}{dt}$ = the rate at which gas moles are consumed per second by the hydrate, A_p = the surface area of each particle, f_{bi} = the fugacity of component i in the bulk liquid, f_{eqi} = the equilibrium fugacity of component i in the liquid at the hydrate interface, K^* = the constant representing the rate of hydrate formation growth, which combines both the rate constants for diffusion (mass transfer) and adsorption (reaction) processes, k_r = the rate constant for the reaction, k_d = the coefficient for mass transfer through the film surrounding the particle and $f_{bi} - f_{eqi}$ = defines the overall driving force.

However, it is suggested that intrinsic kinetics alone may not have a major impact on hydrate formation in real systems, especially in turbulent pipeline flow. Instead, heat and mass transfer are believed to play a more substantial role in determining the rate of hydrate formation, as also indicated in the research work from Englezos et al.

However, it is crucial to acknowledge that the model developed by Englezos et al. in 1987 played a pivotal role by offering a conceptual framework that effectively integrate intrinsic kinetics with factors associated with heat and mass transfer constraints. This approach opened the door for subsequent researchers to utilize Englezos et al.'s model as a fundamental building block for their own models to address situations where the rate-limiting aspects of a process were significantly influenced by heat and mass transfer phenomena.

Heat Transfer Limited Model

Hydrate formation in various processes often faces challenges linked to heat transfer, and having a thorough understanding of these limitations becomes essential, particularly in scenarios where heat transfer may hinder with the formation and growth of hydrates.

In such cases, the effectiveness of hydrate formation processes can be notably impacted by the system's capability to efficiently dissipate heat, given that hydrate formation is an exothermic reaction that releases heat. This means that if the heat transfer within the system is not optimized, it can slow down or hinder the hydrate formation process. To address this challenge and ensure the smooth operation of hydrate formation processes, it is essential to focus on improving heat transfer mechanisms. Numerous models have been developed to elucidate the lateral growth of a hydrate film at the interface between water and gas.

Uchida et al. proposed a heat transfer limited hydrate formation model based on experiments conducted with a water droplet surface submerged in liquid carbon dioxide. In this model, hydrate crystals exclusively form at the leading edge of the hydrate film, and this front region is maintained at the temperature of three-phase equilibrium (water-guest-hydrate). The model assumes steady heat transfer from this front region to both the water and guest fluid. Uchida et al. postulated that the heat extracted from the front region is balanced by the heat generated during hydrate formation, which is an exothermic process. Assuming the balance between the heat transfer away from the film front and the heat released by the hydrate crystal formation at the front, the following equation was proposed [17]:

$$v_f \Delta h_H \rho_H = -k\left(\frac{\partial T}{\partial r}\right)_{rc} \approx -k\left[\frac{T_S - T_{inf}}{r_c}\right] \approx -k\frac{(T_S - T_{inf})}{\delta/2} \tag{6}$$

They established a relationship between the linear growth rate of the hydrate film along the interface (v_f), the hydrate film thickness (δ), and the degree of subcooling ($T_s - T_{inf}$). In the equation above, Δh_H is the latent heat of the hydrate formation and ρ_H is the mole density of the hydrate. The value of δ was determined by fitting the calculated plot of v_f against subcooling to the corresponding experimental data.

Within equation (6), the sole unidentified parameter is the film thickness (δ). Subsequently, this parameter is fine-tuned to achieve the closest match with the experimental data, as shown below.

In Figure 6, it is evident that the model developed by Uchida et al. [17] does not align well with the experimental data. Several potential reasons for this discrepancy can be identified:

a) The estimation of thermal conductivity in the surrounding phases only considered water and neglected the thermal conductivity of CO_2. However, the difference in thermal conductivity between the two is approximately five times in a liquid CO_2-water system, potentially introducing some inaccuracy.

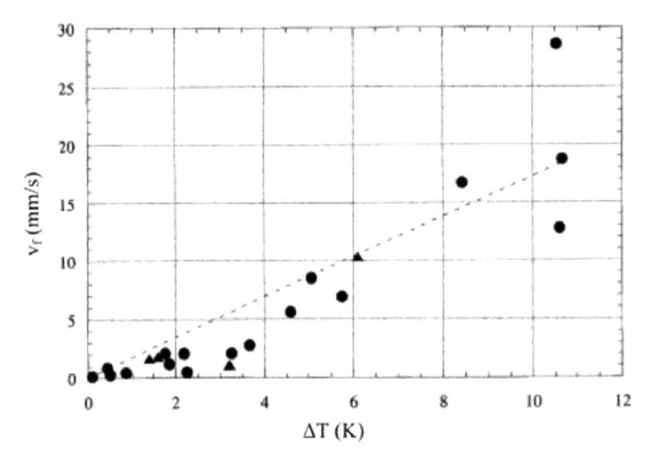

Figure 6. The model [17] (represented by dashed lines) in contrast with the experimental data points, after calibration for a film thickness of 0.13 μm.

b) According to Mori [18], the assumed temperature gradient at the hydrate film front lacks strong physical justification.

c) The model correlates the data of v_f versus ΔT (temperature difference) using a linear regression. However, as depicted in Figure 6, a linear approximation may not be a good fit.

Additionally, Mochizuki and Mori have pointed out a significant issue with Uchida et al. **'s'** model, primarily concerning the formulation of conductive heat transfer from the film front, which lacks physical justification. Conducting direct measurements of the hydrate film thickness under pressure conditions is generally much more challenging than measuring lateral film growth [17, 19]. Consequently, it was acknowledged that the Uchida et al. [7] 's' model offers an approach to estimate the hydrate film thickness [17].

In the model proposed by Mochizuki and Mori, they considered transient two-dimensional conductive heat transfer from the film front to both the water and guest fluid phases, including the hydrate film itself. In this model, it is assumed that the hydrate film exists on the water side of the water-guest fluid interface. The hydrate film is considered homogeneous on a macroscopic scale, while the water and guest phases are assumed to extend infinitely. The general equation for the linear growth rate of the hydrate film along the water and hydrate-former interface, denoted as v_f, can be expressed as below [19]:

$$v_f \Delta h_H \rho_H \delta = -k\left(\frac{\partial T}{\partial r}\right)_{rc} = \int\left(\lambda_h \frac{\partial T}{\partial x_{(x=x_h-)}} - \lambda_w \frac{\partial T}{\partial r_{(x=x_h+)}}\right)dx \tag{7}$$

Here, the variables are defined as follows: δ represents the thickness of the hydrate film, $\frac{\partial T}{\partial x_{(x=x_h-)}}$ and $\frac{\partial T}{\partial r_{(x=x_h+)}}$ denote the temperature gradients on the hydrate side and water side, respectively, at the position $x = x_h$, which is the location of the hydrate-film front, h_H represents the heat of hydrate formation per unit mass of hydrate, λ_h and λ_w represent the thermal conductivity of hydrate and water, respectively.

The above equation describes the rate at which the hydrate film grows along the interface and considers factors such as temperature gradients, heat of hydrate formation, and the thermal conductivity of both hydrate and water. The model assumes that lateral growth of the hydrate film is faster than changes in film thickness, whether thickening or thinning. In this scenario, hydrate crystals only form at the front of the hydrate film, and the temperature at this front region is kept at the three-phase equilibrium temperature.

As a simplification, Freer introduced an equation that encapsulated both kinetic and heat transfer resistance within a single overarching rate constant [20]. This equation is expressed as follows:

$$\Delta h_H \rho_H \frac{dX}{dT} = K(T_{eq} - T_{bulk}) \tag{8}$$

In this equation, K is the total resistance, h is the heat transfer coefficient, and k is the methane hydrate kinetic rate coefficient, Δh_H is the heat of hydrate dissociation, ρ_H is the hydrate density and $\frac{dX}{dT}$ is the rate of the film lateral growth.

$$\frac{1}{K} = \frac{1}{h} + \frac{1}{k} \tag{9}$$

Figure 7 below displays the outcomes obtained from the model proposed by Freer [20] in comparison to the experimental data. The model does not account for any movement of the guest or water phases, and it maintains a balance between the rate of heat removal from the front due to conduction and the rate of heat generation resulting from hydrate formation. It is crucial to highlight that the Freer [20] model goes beyond addressing heat transfer and integrates the kinetics of crystallization. Nevertheless, it is worth noting that this model requires information on both the equilibrium and bulk phase temperatures to make precise predictions regarding hydrate growth in processes where heat transfer is a limiting factor.

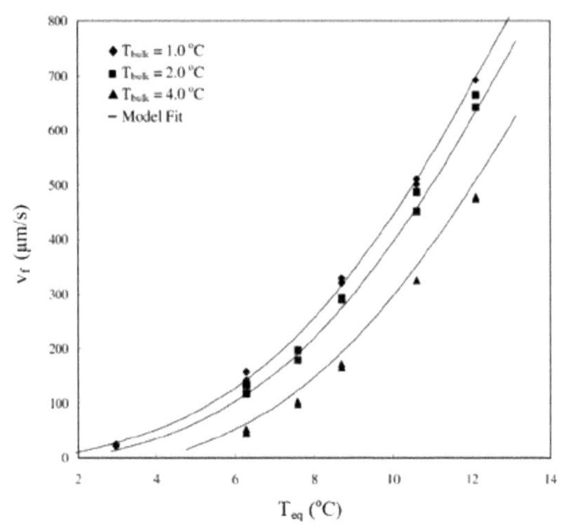

Figure 7. Freer's model compared to the experimental data, with permissions from Freer [20].

Mass Transfer Limited Model

In the context of mass transfer processes, hydrate formation can encounter constraints related to the ability to transfer the hydrate forming molecules to the reaction interface effectively. In such instances, the efficiency of hydrate formation processes can be significantly influenced by the system's capacity to effectively transfer mass. This means that if mass transfer within the system is not optimized, it can slow down or impede the hydrate formation process which can control the rate of hydrate formation. Fundamentally, the mass transfer-limited mechanisms are grounded in the concept of diffusional boundaries, which posits that the rate of change in hydrate growth $\left(\frac{dm}{dt}\right)$ is governed by the diffusion of the hydrate-forming substance from the bulk solution to the equilibrium interface. This can be mathematically expressed as follows:

$$\frac{dm}{dt} = k_d A(C - C_{eq}) \tag{10}$$

In this equation, C and C_{eq} denote the concentrations of a hydrate former in a supersaturated solution and at equilibrium, respectively. A represents the reaction surface area, and k_d stands for the mass transfer coefficient.

This implies that both stages—the diffusion of hydrate forming molecules on the surface and the actual formation of hydrates at the reaction interface—are critical in hydrate formation processes that are constrained by mass transfer phenomena. If we illustrate this concept visually, we can observe a concentration gradient within a boundary layer located on the fluid side of the interface, as depicted below:

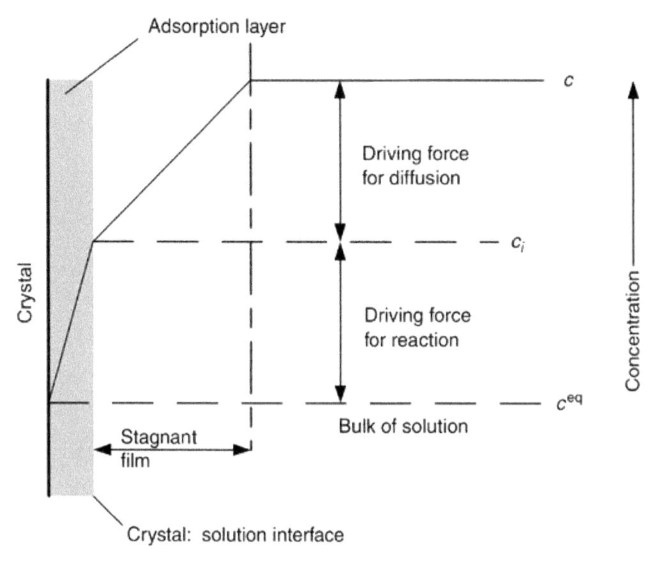

Figure 8. Conceptual model of mass transfer processes from bulk phases to hydrate forming interface, with permission from Sloan and Koh [2].

The driving force for diffusion can be depicted as the difference between the bulk fluid concentration (C) and the interfacial concentration (C_i) within the fluid. Conversely, the driving force for the hydrate-forming reaction can be illustrated by the difference between the equilibrium concentration (C_{eq}) and the interfacial concentration (C_i) within the fluid.

The first equation represents the step concerning the driving force for the diffusion step.

$$\frac{dm}{dt} = k_d A(C - C_i) \tag{11}$$

The second equation represents the reaction step between the hydrate former and water at the reaction interface.

$$\frac{dm}{dt} = k_r A(C_i - C_{eq}) \tag{12}$$

The mass transfer coefficient (k_d) governs hydrate formation when the reaction occurs very quickly compared to diffusion, whereas the reaction coefficient (k_r) dominates hydrate formation when diffusion is much faster than the reaction.

The overall resistance, denoted as (1/K), can be effectively managed by ensuring a low value for either of the individual coefficients. When the reaction occurs at a significantly faster rate compared to diffusion, the mass transfer coefficient (k_d) governs hydrate formation. Conversely, when diffusion significantly outpaces the reaction rate, the reaction coefficient (k_r) takes precedence in controlling hydrate formation. In such scenarios, the overall coefficient K can be approximated by the smaller of the two coefficients, but it is important to note that the concentrations within the driving force remain measurable (C) or computable (C_{eq}), as opposed to being unobservable (C_i).

In this equation, the overall transfer coefficient, denoted as K, is expressed in terms of the diffusion coefficient k_d and the reaction rate constant k_r as:

$$\frac{1}{K} = \frac{1}{k_d} + \frac{1}{k_r} \tag{13}$$

In the Skovborg and Rasmussen model [21], a transformation involving the transfer of the component across the liquid side of the vapour-liquid interface was proposed in the concentration equation. This modification results in the following expression:

$$\frac{dn_i}{dt} = k_L A_{hw}(x_i^{int} - x_i^b) \tag{14}$$

In this context, k_L represents the coefficient responsible for mass transfer across the liquid boundary at the gas-liquid interface, A_{hw} signifies the area of the hydrate former and water interface, while x_i^{int} and x_i^b respectively indicate the mole fractions of hydrate forming component at the interfacial (equilibrium) and bulk levels under the system's prevailing temperature and pressure conditions. In all models pertaining to hydrate growth, the coefficient K is a parameter that is adjusted to conform to experimental kinetic data.

8.8 Model Validation

8.8.1 Model Validation Using Experimental Data

Expanding upon the processes governed by mass and heat transfer limitations, Amiri et al. [22] introduced a multi-stage solid-continuous-solution crystal growth model (MSSCM). This model was developed based on experimental CO_2 gas uptake data from the CO_2/N_2 system as reported by Linga et al. [23]. It is worth mentioning that, in the study, the operating pressure was considerably lower than the thermodynamically required pressure for N_2 hydrate formation. Therefore, the authors assumed that no nitrogen hydrate formation took place during the process [22].

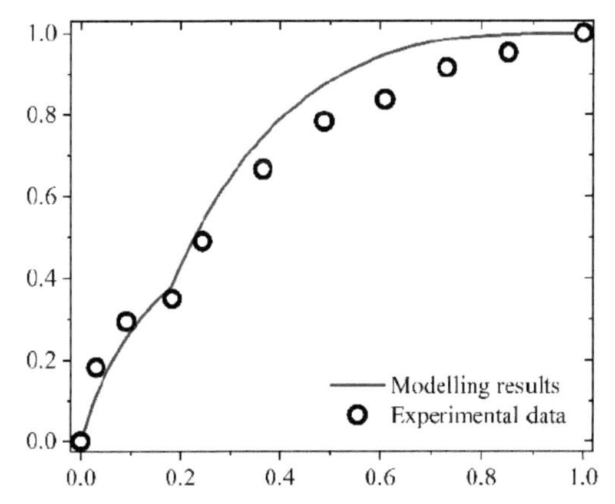

Figure 9. Observed and predicted normalized CO_2 consumed versus normalized time (t/t_{max}.) [22].

In the model, the authors made the following assumptions:

1. The water droplet is initially pure and spherical with a constant radius throughout the process.
2. When the process starts, the reactor pressure and temperature are 3.25 MP and 270 K, respectively.
3. The concentration of CO_2 in the bulk gas was determined using the Peng-Robinson equation of state for CO_2.
4. CO_2 dissolution was assumed to be included in the nucleation processes.

8.8.2 Model Validation Using Process Simulation Software

A hydrate kinetics model called CSMHyK has been developed by the researchers [24] at the Colorado School of Mines, integrated into the OLGA® multiphase simulator. CSMHyK is a transient gas hydrate model to forecast the formation and transportability of gas hydrates within flowlines. CSMHyK is an add-on module designed for the OLGA multiphase flow simulator. This module is designed to forecast the rate of hydrate formation by employing a first-order rate equation based on the thermal driving force. The initial kinetic rate equation, proposed by Vysniauskas and Bishnoi in 1983, was based on a fugacity driving force assumption and did not consider limitations in mass and heat transfer. Subsequently, Turner et al. [25] introduced a more convenient thermal driving force, labelled as ΔT, and accordingly adapted the kinetic equation, as illustrated in the equation below:

$$\frac{dm_{gas}}{dt} = uk_1 e^{\frac{-k_2}{T}} A_s \Delta T \tag{15}$$

In this context, m_{gas} denotes the quantity of gas consumed during the process of hydrate formation, while k_1 and k_2 represent the intrinsic rate constants, with ΔT representing the subcooling. As water droplets undergo a transformation into hydrate particles, the available surface area for hydrate formation, denoted as os, decreases, consequently leading to a decline in the rate at which hydrate forms.

Figure 10 illustrates the cumulative gas injection into a University of Tulsa flow loop during a constant pressure experiment. This section presents a comparison between the model's predictions regarding hydrate formation and the experimental data derived from flow loop experiments. The intrinsic kinetics equation was employed to estimate the rate of hydrate formation, incorporating a correction factor of 0.002. Figure 10 displays the evolution of the hydrate volume fraction over time, obtained from a constant-pressure experiment in the ExxonMobil flow loop using Conroe oil. Hydrate formation began at approximately 0.7 hours, exhibiting an initial high formation rate that slightly decreased within a few minutes. Subsequently, the rate of hydrate formation remained relatively constant for a significant portion of the experiment and gradually decelerated towards the end, coinciding with nearly complete conversion of water into hydrate.

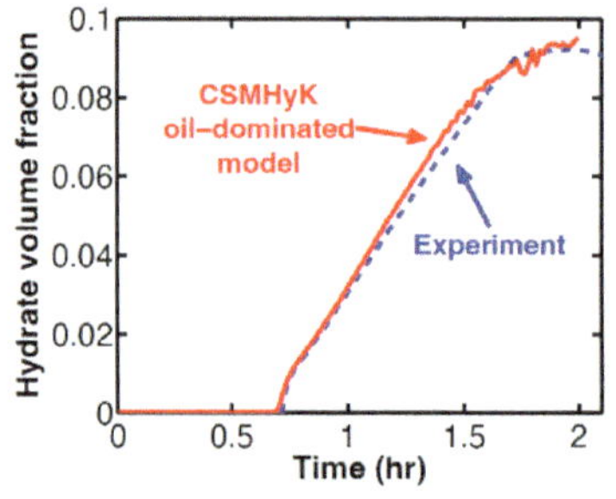

Figure 10. Plot of hydrate volume fraction as the function of time, comparing experimental data and oil-dominated model prediction for an ExxonMobil flow loop constant pressure experiment with Conroe oil, 70% liquid loading, 37% water cut, and 2.3 m/s fluid mixture velocity, taken with permission from PhD thesis of L. Zerpa [24].

8.9 Current Challenges and Opportunities in Hydrate Modelling

Modelling gas hydrates is a complex and challenging task due to the intricate interplay of various physical and chemical processes involved in hydrate formation and dissociation. Some of the key challenges in hydrate modelling efforts include:

Data Availability: Hydrate modelling relies on the availability of high-quality experimental data for validation and parameterization, especially when determining coefficients related to the resistances governing kinetics, heat transfer, or mass transfer rate-limiting steps. Unfortunately, there is often a lack of comprehensive and reliable data, especially when it comes to carbon dioxide hydrate formation, as researchers have predominantly focused on methane or natural gas hydrates for commercial applications. Consequently, accurately predicting the growth of CO_2 hydrates can pose a significant challenge.

Multi-Phase and Multi-Component Systems: In real-world CO_2 capture applications, there are typically multiple gas species involved, which can result in the formation of hydrates within a complex multi-component system. To model these multi-phase, multi-component systems accurately, advanced equations of state and thorough phase equilibrium calculations are necessary. In many cases, simple lab calculations that focus on single-phase behaviour may not provide accurate predictions for growth rates when dealing with the intricate phase behaviour of multi-component systems.

Effect of Impurities: The presence of impurities in both the gas and water phases can have a substantial influence on the nucleation, crystallization, and growth stages of hydrate formation. Modelling the behaviour of these impurities and their effects on hydrate stability is a challenging task and is often overlooked. However, it should be noted that the reaction resistances such as kinetics, mass transfer and heat transfer may change in the presence of impurities and can affect the underlying growth mechanisms.

Uncertainty Quantification: Assessing and quantifying uncertainties in hydrate modelling results is essential for risk assessment and decision-making. Developing methods for robust uncertainty quantification is an ongoing challenge.

Computational Challenges: Performing simulations and modelling hydrate systems can be computationally intensive, especially for large-scale reservoir simulations. Adequate computational resources are needed to tackle complex hydrate modelling problems.

Incorporating Advanced Techniques: Emerging techniques, such as machine learning and data-driven modelling, offer opportunities to enhance hydrate modelling but also present challenges in terms of integration and validation.

It is important to note that research in hydrate modelling is progressing, and great advancements have been made in the last two decades. Researchers are continually working to address these challenges and improve our understanding of gas hydrate behaviour in various contexts.

The potential future directions and advancements in modelling hydrate-based CO_2 capture include:

Enhanced Accuracy in Simulation Models: This is a continual refinement of simulation models to achieve higher accuracy in predicting hydrate formation and growth under various conditions, enabling more precise optimization of CO_2 capture processes.

Incorporating Real-World Complexity: Advancements may involve incorporating real-world complexities such as impurities in the gas stream, fluctuating operational conditions, and the presence of other hydrate-forming substances, making the models more robust and applicable to practical scenarios.

Multi-Scale Modelling: Utilizing multi-scale modelling approaches to bridge the gap between molecular-level interactions and macroscopic system behaviour, offering a more comprehensive understanding of the hydrate formation process.

Machine Learning Integration: Incorporating machine learning and data-driven techniques to improve model accuracy, adaptability, and the ability to learn from real-time data, leading to more adaptive and efficient CO_2 capture systems.

These advancements in modelling hydrate-based CO_2 capture hold the potential to make this technology more efficient, cost-effective, and environmentally friendly, contributing to the ongoing efforts to mitigate climate change.

8.10 Conclusions

By exploring the methods for modelling and simulating CO_2 capture utilizing hydrate technology, this chapter aims to provide researchers, engineers, and policymakers with valuable insights into the theoretical and computational aspects of this promising approach. Overall, it serves as a valuable resource for understanding the complexities, opportunities, and challenges associated with modelling and simulating hydrate-based CO_2 capture processes.

References

[1] Englezos, P., Kalogerakis, N., Dholabhai, P. D. and Bishnoi, P. R. 1987. Kinetics of gas hydrate formation from mixtures of methane and ethane. Chemical Engineering Science 42(11): 2659–2666.

[2] Sloan Jr, E. D. and Koh, C. A. 2007. Clathrate Hydrates of Natural Gases. CRC Press.

[3] Khan, M. N., Rovetto, L. J., Peters, C. J., Sloan, E. D., Sum, A. K. and Koh, C. A. 2015. Effect of hydrogen-to-methane concentration ratio on the phase equilibria of quaternary hydrate systems. Journal of Chemical & Engineering Data 60(2): 418–423.

[4] Maeda, N. 2019. Nucleation curve of carbon dioxide hydrate. Energy Procedia 158: 5928–5933.
[5] Christiansen, R. L. and Sloan, E. D. 1995. In Proceedings of the 74th Annual Convention of the Gas Processors Association, San Antonio, TX, 15 March 1995.
[6] Kashchiev, D. and Firoozabadi, A. 2002. Nucleation of gas hydrates. Journal of Crystal Growth 243(3-4): 476–489.
[7] Anklam, M. R. and Firoozabadi, A. 2004. Driving force and composition for multicomponent gas hydrate nucleation from supersaturated aqueous solutions. The Journal of Chemical Physics 121(23): 11867–11875.
[8] Abay, H. K. and Svartaas, T. M. 2011. Multicomponent gas hydrate nucleation: the effect of the cooling rate and composition. Energy & Fuels 25(1): 42–51.
[9] Walsh, M. R., Koh, C. A., Sloan, E. D., Sum, A. K. and Wu, D. T. 2009. Microsecond simulations of spontaneous methane hydrate nucleation and growth. Science 326(5956): 1095–1098.
[10] Radhakrishnan, R. and Trout, B. L. 2002. A new approach for studying nucleation phenomena using molecular simulations: Application to CO_2 hydrate clathrates. The Journal of Chemical Physics 117(4): 1786–1796.
[11] Jacobson, L. C., Hujo, W. and Molinero, V. 2010. Nucleation pathways of clathrate hydrates: Effect of guest size and solubility. The Journal of Physical Chemistry B 114(43): 13796–13807.
[12] Jacobson, L. C., Hujo, W. and Molinero, V. 2010. Amorphous precursors in the nucleation of clathrate hydrates. Journal of the American Chemical Society 132(33): 11806–11811.
[13] Bi, Y., Porras, A. and Li, T. 2016. Free energy landscape and molecular pathways of gas hydrate nucleation. The Journal of Chemical Physics 145(21).
[14] Barnes, B. C., Knott, B. C., Beckham, G. T., Wu, D. T. and Sum, A. K. 2014. Reaction coordinate of incipient methane clathrate hydrate nucleation. The Journal of Physical Chemistry B 118(46): 13236–13243.
[15] Kim, H. C., Bishnoi, P. R., Heidemann, R. A. and Rizvi, S. S. 1987. Kinetics of methane hydrate decomposition. Chemical Engineering Science 42(7): 1645–1653.
[16] Moridis, G. J., Kowalsky, M. B. and Pruess, K. 2008. TOUGH+ HYDRATE v1. 0 User's Manual: a code for the simulation of system behaviour in hydrate-bearing porous media. Report LBNL-149E. Lawrence Berkeley National Laboratory, Berkeley, CA.
[17] Uchida, T., Ebinuma, T., Kawabata, J. I. and Narita, H. 1999. Microscopic observations of formation processes of clathrate-hydrate films at an interface between water and carbon dioxide. Journal of Crystal Growth 204(3): 348–356.
[18] Mori, Y. H. 2001. Estimating the thickness of hydrate films from their lateral growth rates: application of a simplified heat transfer model. Journal of Crystal Growth 223(1-2): 206–212.
[19] Mochizuki, T. and Mori, Y. H. 2006. Clathrate-hydrate film growth along water/hydrate-former phase boundaries—numerical heat-transfer study. Journal of Crystal Growth 290(2): 642–652.
[20] Freer, E. M. 2000. Methane Hydrate Formation Kinetics. Chemical Engineering. Golden, Colorado School of Mines. Masters.
[21] Skovborg, P. and Rasmussen, P. 1994. A mass transport limited model for the growth of methane and ethane gas hydrates. Chemical Engineering Science 49(8): 1131–1143.
[22] Dashti, H., Thomas, D. and Amiri, A. 2019. Modeling of hydrate-based CO_2 capture with nucleation stage and induction time prediction capability. Journal of Cleaner Production 231: 805–816.
[23] Linga, P., Kumar, R. and Englezos, P. 2007. Gas hydrate formation from hydrogen/carbon dioxide and nitrogen/carbon dioxide gas mixtures. Chemical Engineering Science 62(16): 4268–4276.

[24] Zerpa, L. E., Sloan, E. D., Sum, A. K. and Koh, C. A. 2012. Overview of CSMHyK: A transient hydrate formation model. Journal of Petroleum Science and Engineering 98: 122–129.

[25] Turner, D., Boxall, J., Yang, S., Kleehammer, D. M., Koh, C. A., Miller, K. T. et al. (2005, June). Development of a hydrate kinetic model and its incorporation into the OLGA2000® transient multiphase flow simulator. pp. 1231–1240. In 5th International Conference on Gas Hydrates (Vol. 4018). Trondheim, Norway: Tapir Academic Press.

CHAPTER 9

Leveraging Machine Learning and Artificial Intelligence for Enhanced Carbon Capture and Storage (CCS)

Jai Krishna Sahith[1,*] and *Bhajan Lal*[2]

9.1 Introduction

The global community is confronting an unprecedented challenge—climate change—propelled by the relentless emission of greenhouse gases into the Earth's atmosphere. The urgency of mitigating the adverse impacts of climate change has driven the quest for innovative solutions to reduce and ultimately eliminate these emissions. Carbon Capture and Storage (CCS) has emerged as a promising technology to curtail carbon dioxide (CO_2) emissions from industrial processes and energy generation [1]. By capturing CO_2 emissions at the source and securely storing them underground, CCS has the potential to significantly reduce the concentration of greenhouse gases in the atmosphere.

However, the effectiveness and efficiency of CCS technologies are pivotal to their widespread adoption and success in achieving emissions reduction targets. This is where the fusion of Machine Learning (ML) and Artificial Intelligence (AI) with CCS becomes a game-changing proposition [2]. In recent years, these advanced technologies have catalyzed a paradigm shift in the way we approach and optimize complex processes, and their application to CCS holds great promise in making

[1] School of Chemical & Bioprocess Engineering, University College Dublin, Ireland.
[2] Chemical Engineering Department, Center for Carbon Capture, Storage and Utilization (CCUS), Institute of Sustainable Energy, Universiti Teknologi PETRONAS, Bandar Seri Iskandar, Perak, Malaysia.
* Corresponding author: jaikrishnasahith1906@gmail.com

this vital climate mitigation strategy more effective, economically viable, and environmentally sustainable.

The intersection of ML and AI with CCS is a frontier of research and innovation, as it promises to address critical challenges such as enhancing capture efficiency, optimizing transport and injection of captured CO_2, monitoring and maintaining the integrity of storage sites, and ensuring the long-term viability of CCS projects [3]. ML and AI can harness the power of predictive analytics, real-time data analysis, and autonomous decision-making, thus enabling CCS to reach its full potential [4].

This research article explores the burgeoning field of applying ML and AI to CCS, delving into the myriad ways in which these technologies are revolutionizing the landscape of carbon capture and storage. We will elucidate their application in improving the design and operation of CCS facilities, ensuring secure and leak-free storage, and optimizing the overall CCS lifecycle [5]. Through a comprehensive review of existing studies and a critical assessment of the state of the art, we aim to shed light on the immense possibilities and ongoing challenges of merging ML and AI with CCS technology [6].

To navigate this fascinating realm, we will embark on a journey that explores data-driven insights and predictive models that can significantly enhance CCS systems. Additionally, we will consider the ethical and regulatory aspects associated with the integration of advanced technologies in the energy and environmental sectors [7]. It is important to recognize that the success of CCS, bolstered by ML and AI, has the potential to reshape our global approach to climate change mitigation [8, 9].

In the chapters that follow, we will navigate the intricacies of this promising field, drawing inspiration from pioneering research, and shedding light on the emerging trends and challenges. Together, we will embark on a transformative expedition, harnessing the power of Machine Learning and Artificial Intelligence to turn the tide in the fight against climate change, as we unlock the true potential of Carbon Capture and Storage.

9.2 Carbon Capture or Separation

Carbon dioxide (CO_2) emissions, stemming primarily from the combustion of fossil fuels and industrial processes, have elevated our planet to the precipice of a global climate crisis. In 2018, the Earth's atmospheric CO_2 concentration soared to a record 416 parts per million (ppm), an unparalleled level in at least 800,000 years [10]. The repercussions of this surge manifest in the form of ascending global temperatures, intensified weather events, and the looming spectre of ecological instability.

In response to this alarming reality, the world's collective determination to combat climate change has never been more resolute. At its core lies the imperative to advance innovative CO_2 capture and separation technologies, which are pivotal components in the complex puzzle of greenhouse gas mitigation [11].

The stakes are undeniably high, as underscored by the Intergovernmental Panel on Climate Change (IPCC) in their 2021 report, emphasizing the critical importance of limiting global warming to within 1.5°C above pre-industrial levels to avert catastrophic climate impacts [12]. Achieving this ambitious goal mandates an immediate and substantial reduction in CO_2 emissions. While renewable energy sources and enhanced efficiency measures are integral, the linchpin to addressing emissions from sectors that defy easy decarbonization, such as heavy industry and specific forms of transportation, is unequivocally carbon capture and separation technologies [13].

This chapter section goes on an exhaustive exploration into the advances, obstacles, and future potential of CO_2 capture and separation technologies. It is our assertion that these technologies represent the cornerstone in the global endeavour to curtail emissions and propel the transition towards a sustainable, low-carbon economy.

Anderson et al. delved into Open Framework Materials (OFMs), with a specific focus on Metal-Organic Frameworks (MOFs) and their potential as a game-changer in the realm of carbon dioxide (CO_2) capture [14]. MOFs, with their customizable nanopores, offer a unique opportunity to strategically position optimal adsorption sites within their structures, paving the way for selective CO_2 physisorption—a key advancement in energy-efficient CO_2 capture technology. The crux of the matter lies in deciphering which features of MOFs are pivotal for enhancing CO_2 capture capabilities. With thousands of MOFs synthesized to date, each with a multitude of adjustable properties, this task can be daunting. This research employed a multiscale approach, integrating Density Functional Theory (DFT), Grand Canonical Monte Carlo (GCMC), and Machine Learning (ML) to unravel the complex relationship between various pore chemical and topological features and their impact on CO_2 capture metrics within MOFs.

To conduct a comprehensive exploration of MOF structure-space, the team used computational synthesis methods to create a diverse set of MOFs by combining 16 topologies and 13 functionalized molecular building blocks. They subsequently simulated the adsorption of pure CO_2 as well as CO_2 in mixtures with H_2 and N_2 for the resulting MOFs and their derivatives, ultimately calculating CO_2 capture metrics. One intriguing finding of the study was that functionalization with various chemistries, including hydroxyl, thiol, cyano, amino, or nitro groups, often enhanced the CO_2 capture metrics of parent MOFs. However, the effectiveness of this strategy was notably influenced by the pore topology. To remove any human bias, the research team employed machine learning algorithms, including decision trees, to predict the impact of functionalization on CO_2 capture metrics, as well as to determine the relative importance of different pore chemical and structural/topological factors in the CO_2 capture capabilities of MOFs.

In essence, this research provides an exciting perspective on the pivotal role of MOFs in advancing CO_2 capture technology. By meticulously examining the interplay of pore chemistry and topology, and by leveraging machine learning, the study sheds light on the intricacies of these materials' capabilities. The findings have the potential to guide the future design and development of MOFs tailored for optimal CO_2 capture, thus contributing to the global effort to combat climate change through more efficient carbon capture strategies.

Dureckova et al. opened a compelling chapter in the field of precombustion carbon capture by leveraging robust machine learning models and Quantitative Structure-Property Relationships (QSPR) to predict critical parameters for the efficient adsorption of CO_2 in Metal-Organic Frameworks (MOFs) [15]. Their focus extends to both CO_2 working capacity and CO_2/H_2 selectivity, crucial factors in gas separation applications for carbon capture. What sets this work apart is the vast diversity of the MOF structures explored. The researchers assembled a database featuring a staggering 358,400 MOFs with 1,166 network topologies. This diversity is a significant leap forward, far surpassing the limited range of topologies used in prior research. The dataset's richness empowers the development of machine learning models that can unlock the potential of a broader spectrum of MOFs.

The heart of their approach lies in the gradient-boosted trees regression method, which enables the use of 80% of the expansive database for training, leaving the remaining 20% for validation and testing. The initial models are crafted using geometric descriptors of MOFs, including parameters such as gravimetric surface area and void fraction. Beyond this, the researchers introduced models that consider the chemical features of MOFs, incorporating Atomic Property-weighted Radial Distribution Functions (AP-RDFs) with innovative normalization techniques to accommodate the database's size diversity. The results are impressive. The models constructed from a combination of six geometric descriptors and three AP-RDF descriptors achieve remarkable accuracy in predicting CO_2 working capacities ($R^2 = 0.944$) and CO_2/H_2 selectivities ($R^2 = 0.872$). However, what truly stands out is the models' capability to identify the top 1,000 high-performing MOFs within an extensive pool of 3,000 or 5,000 MOFs. This significantly accelerates the screening process for top-performing MOFs, a critical development in the quest for efficient precombustion carbon capture.

In summary, this research demonstrates the power of machine learning and QSPR modelling in navigating the intricate landscape of MOF diversity. By developing models that can recognize high-performing MOFs for carbon capture applications, even amidst a vast array of structural and chemical variations, the study not only accelerates the screening process but also enhances our capacity to harness MOFs for sustainable and effective precombustion carbon capture. It marks a significant step toward a more efficient and sustainable approach to mitigating carbon emissions.

Leperi et al. presented a compelling contribution to the pivotal realm of carbon capture technologies, acknowledging their integral role in shaping the global energy landscape [16]. In an era where fossil fuels continue to dominate the energy mix, the need for efficient and economically viable carbon capture solutions has never been more pronounced. Pressure Swing Adsorption (PSA) stands out as a promising alternative, with its minimal energy requirements, making it a potent candidate for carbon capture applications. Yet, the challenge lies in designing the optimal PSA cycle for a given adsorbent material, a critical step towards commercial competitiveness.

In this work, the authors proposed and put to test a model reduction-based approach, offering a systematic path to generate low-order representations of complex PSA models. These reduced-order models are a product of training Artificial Neural Networks (ANNs) using data derived from full Partial Differential Algebraic Equation (PDAE) model simulations. The core contribution of this research is the development of surrogate models for every step in PSA cycles, spanning pressurization, adsorption, and depressurization phases, for both concurrent and counter-current operations. To validate their approach, the researchers employed three distinct PSA cycles, namely the three-step, Skarstrom, and five-step cycles, all intended for post combustion carbon capture. Two adsorbents, Ni-MOF-74, and zeolite 13X, were selected for evaluating the surrogate models under optimized cycle conditions.

The results are compelling, with excellent agreement observed between the ANN models and the PDAE simulations. The average mean square errors, with respect to dimensionless state variables, were found to be remarkably low, indicating a high degree of accuracy in the surrogate models. Furthermore, the highest relative error concerning CO_2 purity and recovery was a mere 1.42%, affirming the feasibility of implementing machine learning techniques to create PSA surrogate models. These models offer a promising avenue for optimization in synthesizing PSA cycles, potentially propelling the commercial viability of carbon capture through pressure swing adsorption.

In sum, this research is a testament to the power of AI and machine learning in streamlining complex processes in pursuit of sustainable energy solutions. By harnessing the capabilities of surrogate models, the study not only advances the field of carbon capture but also underscores the potential of AI to drive innovation in the broader context of renewable and clean energy technologies.

The intersection of machine learning (ML) and carbon capture is rapidly becoming a cornerstone in the global quest to mitigate climate change. The research work by Rahimi et al. underscored the transformative role of ML in revolutionizing the carbon capture landscape [17]. Carbon capture is an essential interim technology for curbing climate change, and ML has emerged as a powerful ally in enhancing this process. This perspective delves into the multitude of ways ML is making its mark in carbon capture, spanning from absorption to adsorption-based approaches, and operating at scales from the molecular to the process level.

For absorption, ML is deployed to predict the thermodynamic properties of absorbents, significantly advancing our understanding of this crucial aspect of carbon capture. Meanwhile, in the realm of adsorption processes, ML techniques offer a promising path to explore numerous options, enabling the identification of the most cost-effective process scheme. This involves selecting the right solid adsorbent and designing an optimal process configuration, thereby boosting the overall efficiency of carbon capture. The perspective also highlights the inherent strengths of ML while acknowledging the associated risks. It emphasizes the significance of obtaining relevant features to train ML models effectively, a critical step in ensuring their accuracy and reliability. Moreover, the future of ML in carbon capture processes appears promising. As this technology continues to evolve, it holds the potential to catalyze breakthroughs in addressing climate change. By systematically analyzing, predicting, and optimizing the capture of carbon dioxide, ML not only improves the efficiency of current carbon capture methods but also opens doors to innovative and sustainable solutions that can have a profound impact on our collective efforts to combat climate change.

In summary, this perspective elucidates how the fusion of ML and carbon capture is charting a course towards more intelligent, effective, and environmentally friendly solutions. It underscores the evolving role of technology in addressing one of the most pressing challenges of our time, the mitigation of climate change.

The review article by Liu et al. serves as a window into the future of research in the domain of solvent development for post-combustion carbon dioxide (CO_2) capture [18]. While the last five years have witnessed a dynamic evolution in this field, this paper provides a comprehensive synthesis of research spanning from 2015 to 2020, illuminating significant advances and promising directions.

The research areas explored in this review cover a broad spectrum of topics, offering a glimpse into the multifaceted landscape of CO_2 capture. These areas encompass:

1. **Solvent Chemistry:** Investigating aspects such as solubility, reaction kinetics, Nuclear Magnetic Resonance (NMR) analysis for ion

speciation, and the exploration of alternative solvents. Understanding the chemical dynamics of solvents is pivotal in enhancing their efficiency in capturing CO_2.

2. **Development of Amine Solvents:** Focusing on the development of catalysts and the enhancement of existing solvents through catalyst integration. This avenue represents a significant stride toward optimizing the performance of amine solvents.
3. **Electrochemically Mediated Amine Regeneration:** Introducing a novel regeneration process that holds the promise of improving the sustainability and effectiveness of CO_2 capture methods.
4. **Applications of Artificial Intelligence (AI):** Embracing the power of AI technologies to revolutionize the CO_2 capture process. AI is increasingly recognized as a transformative force in optimizing and streamlining complex operations.

This review serves as an invaluable update to the evolving landscape of CO_2 capture research. It not only highlights the remarkable progress made but also offers a critical perspective on the potential paths for future investigations. The integration of AI into CO_2 capture represents a particularly intriguing and forward-looking dimension, emphasizing the significance of technology in addressing contemporary environmental challenges. In essence, this review article encapsulates the spirit of continuous innovation and collaboration in the pursuit of more efficient and sustainable carbon dioxide capture methods. It sheds light on how research in this critical field has evolved and sets the stage for a future where advanced technologies, chemistry, and innovative processes converge to combat climate change effectively.

In the ongoing battle against climate change, Carbon Capture, Utilization, and Storage (CCUS) has emerged as a critical solution to help achieve the ambitious targets outlined in the Paris Agreement. As we seek ways to mitigate the devastating impacts of climate change, leveraging well-established CCUS technologies while continuously improving them becomes paramount. Enter machine learning (ML), a transformative tool that holds the promise of accelerating and enhancing the CCUS value chain.

Yan et al. state-of-the-art review sheds light on the pivotal role of ML in the CCUS landscape [19]. ML, encompassing a suite of advanced statistical tools and algorithms, offers the capacity to classify, predict, optimize, and cluster data, thereby revolutionizing every facet of CCUS. The review takes us on a journey through the key stages of the CCUS value chain, encompassing CO_2 capture, transport, utilization, and storage. It is within each of these stages that ML emerges as a leading force in expanding our knowledge and capabilities. ML's ability to analyze vast datasets, predict outcomes, and optimize processes presents a powerful opportunity to make CCUS more efficient, cost-effective, and sustainable.

The value of this review extends beyond its exploration of the current landscape. It culminates in a set of recommendations for future work and research that can harness the full potential of ML in CCUS. This forward-looking perspective emphasizes the need for ongoing collaboration and innovation, emphasizing that the integration of ML into CCUS is not just an option but a necessity in the global quest to combat climate change. In summary, this review highlights the transformative power of machine learning in advancing CCUS technologies, which are central to achieving our climate goals. It highlights the path forward, where the fusion of technology and environmental stewardship paves the way for a more sustainable and resilient future. The incorporation of ML not only accelerates CCUS but also ensures that it evolves to meet the challenges of the 21st century.

Fathalian et al. highlighted an exciting frontier in the field of carbon dioxide (CO_2) capture using machine learning [20]. As the world grapples with the challenge of mitigating climate change, the development of intelligent prediction models based on machine learning offers a promising avenue for advancing CO_2 capture technology. The study centres around the utilization of graphene oxide (GO) as a foundation for novel and efficient adsorbents to capture CO_2. GO, derived from various forms of solid biomass, holds immense potential for CO_2 adsorption applications. The research extracts crucial data from 17 articles concerning GO-based solid sorbents. Parameters such as specific surface area, pore volume, temperature, and pressure are considered as input, while CO_2 uptake capacity becomes the model's response variable.

The research explores a range of machine learning models, including support vector machine, gradient boosting, random forest, artificial neural network (ANN) using multilayer perceptron (MLP) and Radial Basis Function (RBF), extra trees regressor, and extreme gradient boosting. Among these, the ANN based on the MLP method was found to be superior, achieving an impressive R-squared value of greater than 0.99. The value of this research extends beyond the development of accurate prediction models. It underscores the potential for significantly reducing the need for extensive experimental efforts in assessing CO_2 removal efficiency. By relying on the predictive capabilities of these machine learning models, researchers can efficiently identify and select the most effective adsorbents for CO_2 capture applications, while saving valuable time.

Furthermore, the study delves into three-dimensional diagrams to visualize CO_2 uptake dependency on key parameters, providing deeper insights into the CO_2 adsorption process. It also offers a peek into the weight and bias matrices of the MLP network, facilitating further refinement of

CO_2 adsorption process design. In essence, this research aligns with the broader movement towards more efficient, environmentally friendly manufacturing. By harnessing the power of machine learning, it not only streamlines the development of porous GO for CO_2 separation but also aids in screening adsorbents to make manufacturing cleaner and more sustainable. The study is a testament to the potential of artificial intelligence in advancing environmental solutions and creating a more sustainable future.

Irfan et al. explored the application of Artificial Neural Network (ANN) modelling for carbon dioxide (CO_2) adsorption on various types of Marcellus shale samples [21]. This study delves into eight distinct shale geometries, examining their CO_2 adsorption characteristics under specific conditions. The research investigates CO_2 adsorption at 298 K and pressures up to 50 bar, employing a gravimetric technique and a magnetic suspension balance. The focal points of the ANN modelling endeavour encompass three primary objectives:

1. **Impact of Various Training Algorithms:** The research scrutinizes the influence of different training algorithms on the ANN model, shedding light on their effectiveness in capturing the intricate relationship between CO_2 adsorption and shale properties.
2. **Various Data Initiation Points:** By analyzing different data initiation points, the study aims to uncover how the starting point of data collection impacts the ANN model's performance, providing valuable insights into data selection strategies.
3. **Altered Training/Validating Ratios and Neuron Numbers:** The study explores how changing the ratios between training and validating data, as well as the number of neurons in the ANN model, affects its predictive capacity. This investigation is critical for optimizing model performance.

The outcomes of this research promise to offer valuable insights into the nuanced interplay of these studied parameters within the context of ANN modelling and training algorithms. This knowledge not only enriches our understanding of CO_2 adsorption on Marcellus shale but also holds potential significance for industries involved in unconventional resource exploration, enhanced gas, and oil recovery applications, and beyond. In an era where artificial intelligence is becoming increasingly integral to scientific research and industrial applications, the findings of this study open doors to enhanced modelling and assessment capabilities. As industries seek to leverage artificial intelligence for more efficient and informed decision-making, this research sets the stage for optimized and data-driven approaches in the exploration of unconventional resources, ultimately contributing to more sustainable and effective energy practices.

In the dynamic landscape of carbon capture, storage, transportation, and utilization (CCSTU) processes, Gupta and Li presented a paper that offers a panoramic view of these processes and the transformative role that machine learning (ML) can play in their advancement [22]. The paper explored diverse CO_2 capture and separation technologies (CCT), encompassing absorption, adsorption, membranes, chemical looping, pyrogenic carbon capture and storage (PyCCS), hydrates, and mineral sequestration. The authors provided a comprehensive review of these technologies, highlighting their potential and current state of development. A noteworthy aspect of this research is its classification of hybrid processes, where multiple methods synergistically combine for both CO_2 capture and utilization. Such integration represents a promising avenue for making CCSTU more efficient and effective.

ML methods, as discussed in the paper, are introduced as a key enabler for enhancing the efficiency of CCSTU. ML's application in process optimization and the design of new materials offers the potential to significantly improve the performance and sustainability of CCSTU technologies. The paper does not just stop at a review; it offers a vision for the future. It outlines recommendations for further research, emphasizing the need to consider the carbon impact of AI-driven models. It also underscores the importance of fostering collaboration between ML experts, experimentalists, and computational modellers, advocating for a holistic approach that encourages cooperation over competition in the rapid commercialization of low-carbon technologies. Furthermore, the paper provides examples of databases that can serve as valuable resources for future research, facilitating data-driven exploration and innovation in the field of CCSTU. In essence, this paper encapsulates the evolving spirit of interdisciplinary collaboration and technological innovation in the quest to address climate change. It showcases how CCSTU processes, coupled with the capabilities of machine learning, are positioned to be a pivotal force in realizing low-carbon and sustainable solutions for a more environmentally responsible future.

Narimani's thesis delves into an increasingly critical challenge in the context of mitigating climate change—the efficient capture of carbon dioxide (CO_2) [23]. With rising global temperatures posing a significant threat, efforts to combat this phenomenon have become a top priority. One major source of the temperature increase is carbon emissions, primarily in the form of greenhouse gases. Thus, the capture of CO_2 has emerged as a key area of research focus in recent years. Among the various methods for carbon capture, post-combustion stands out as a common and practical approach, particularly for existing plants. However, it is not without its challenges. Solvent degradation is a central issue within post-combustion carbon capture, with the cost of addressing this problem accounting for a substantial portion of the total operational expenses. Narimani's research focuses on unravelling the complex behaviour of solvent degradation in carbon capture plants by employing machine learning methods. Given the intricate nature of this phenomenon and the absence of traditional methods or software to simulate solvent degradation effectively, machine learning emerges as a promising avenue for predicting and understanding this problem.

The study explores various machine learning methods, with particular emphasis on Artificial Neural Network (ANN), Random Forest, and Support Vector Regression. Leveraging lab and online data from the Technology Centre Mongstad, the research applies feature selection techniques to enhance model performance. These techniques help reduce the number of features, making the models more efficient and interpretable. The results are promising. The study showcases that ANN and Random Forest are the most effective models for predicting solvent degradation behaviour in carbon capture plants. The high R2 scores, often exceeding 0.90 for a substantial dataset from Technology Centre Mongstad, emphasize the accuracy and reliability of these models. In sum, Sam Narimani's research offers a vital contribution to the field of carbon capture. By harnessing machine learning, it provides a pathway to predict and mitigate solvent degradation, a key challenge in enhancing the performance and sustainability of carbon capture plants. The findings hold the potential to significantly reduce the environmental impact of carbon capture, making it a more efficient and cost-effective solution in our collective efforts to combat climate change.

The review conducted by Priya et al. delves into an urgent and pressing issue in today's world: carbon capture, a critical aspect of the global response to climate change [24]. Carbon emissions pose a significant threat to the environment, and addressing this challenge necessitates innovative approaches. In this context, digital technologies, including data pooling and artificial intelligence (AI), have emerged as powerful allies in the quest for environmental sustainability. The study's central theme revolves around the integration of AI into carbon capture processes, marking a pivotal moment in our fight against climate change. Priya and her team explored various techniques, including machine learning (ML), Deep Learning (DL), and hybrid methodologies, and their application in the realm of carbon capture. These AI techniques offer a promising pathway to enhance the efficiency and efficacy of carbon capture technologies.

The research also delves into the role of AI tools, frameworks, and mathematical models. These elements play a crucial role in not only understanding the complex dynamics of carbon capture but also in optimizing these processes for maximum environmental benefit. A notable aspect of this study is the exploration of the patent landscape where AI intersects with carbon capture. This insight sheds light on the innovations and intellectual property emerging in this field, providing a glimpse into the rapidly evolving landscape of AI-assisted carbon capture technologies. In essence, this review contributes significantly to the ongoing efforts to mitigate climate change and align with Sustainable Development Goal 13—Climate Action. By providing a comprehensive overview of the integration of AI in carbon capture, the study equips researchers and stakeholders with the knowledge and perspective needed to harness the potential of AI in the battle against climate change. It underlines the growing importance of AI in creating innovative and sustainable solutions and emphasizes the collaborative effort required to address one of the most significant challenges of our time.

Hosseinpour et al. offered a comprehensive review of an innovative approach to tackling one of the most pressing environmental challenges of our time: carbon capture and storage (CCS). The vast consumption of fossil fuels in various human activities has led to a significant surge in anthropogenic carbon dioxide (CO_2) emissions, resulting in far-reaching and harmful consequences for our planet. Among CCS technologies, absorption-based Post-combustion Carbon Capture (PCC) stands out as a well-established and operational approach. However, the processes of modelling and optimizing PCC, from fine-tuning operating conditions to managing solvents and equipment configurations, have proven to be time-consuming and computationally expensive. This is where the transformative power of machine learning (ML) enters the picture. ML has gained remarkable traction as a tool that can conduct complex computations, enabling the training of algorithms to perform tasks with extraordinary precision, tasks that were previously challenging to accomplish with conventional tools. ML techniques, encompassing classification, prediction, clustering, ranking, and data optimization, offer an efficient and cost-effective approach to revolutionizing the field of PCC [25].

The review in question delves into recent research progress that applies ML methods to PCC absorption-based technologies. It categorizes the various ML methods, offering insights into their limitations, availability, advantages, and disadvantages. This categorization serves as a practical guide for researchers and practitioners, facilitating a more informed approach to choosing and implementing ML techniques in PCC. In closing, the authors present a forward-looking perspective. They proposed a roadmap for community efforts, outlining potential pathways and future research areas for further advancing the application of ML in PCC absorption-based technologies. This roadmap signals a collective commitment to harnessing the full potential of machine learning in addressing climate change and underscores the importance of interdisciplinary collaboration in finding innovative solutions to one of the most significant environmental challenges of our time.

Zhang et al. explored an innovative approach to enhance hydrocarbon production and tackle climate change simultaneously through carbon capture utilization and storage (CCUS) in unconventional formations. The success of such projects hinges on a crucial factor: shale wettability [26].

This study employs multiple advanced machine learning (ML) techniques, including multilayer perceptron (MLP) and Radial Basis Function Neural Networks (RBFNN), to evaluate shale wettability. The evaluation relies on five key features, which encompass formation pressure, temperature, salinity, Total Organic Carbon (TOC), and theta zero. The dataset for this analysis is derived from 229 datasets of contact angles in three different states: shale/oil/brine, shale/CO_2/brine, and shale/CH_4/brine systems. To optimize the ML models, five algorithms are employed for tuning MLP, and three optimization algorithms are used to fine-tune the RBFNN computing framework.

The results of this research are impressive. The RBFNN-MVO model emerges as the most effective, achieving the best predictive accuracy with a minimal Root Mean Square Error (RMSE) value and an exceptionally high R^2 value. The model demonstrates the potential for highly precise predictions, with an RMSE of 0.113 and an R^2 of 0.999993. A sensitivity analysis sheds light on the key features that significantly influence shale wettability. These features include theta zero, TOC, pressure, temperature, and salinity. The identification of these influential factors is pivotal in understanding and optimizing shale wettability.

In essence, this research showcases the power of advanced machine learning techniques in evaluating shale wettability for CCUS initiatives and cleaner production. By leveraging these techniques, it offers a pathway toward more efficient and sustainable hydrocarbon production while making substantial strides in mitigating climate change. It underscores the vital role that interdisciplinary approaches and innovative solutions can play in addressing complex environmental challenges.

Aliyon et al. addressed a significant challenge in the realm of carbon capture, specifically focusing on Absorption-based Carbon Capture (ACC) [27]. While ACC is a mature technology with promise for mitigating CO_2 emissions, it is inherently energy-intensive, necessitating substantial heating and cooling utility consumption, which, in turn, drives up operational costs. To assess the technical and financial feasibility of various ACC process designs, reliable and rapid estimation of these utility consumptions is indispensable. The research objective is to harness the power of artificial intelligence (AI) to predict the utility consumptions of different ACC process designs. The primary target of this prediction is the specific reboiler duty, a crucial energy component of the plant and a major source of operational costs in ACC processes.

The study demonstrates that AI models achieve impressive accuracy, with errors ranging from 0.4 to 3.6% for different scenarios, whether utilizing limited or extensive training data. This high level of precision equips researchers and practitioners with a valuable tool for assessing and optimizing the energy efficiency of ACC plants. Furthermore, the research explores explainable artificial intelligence, shedding light on the contribution of each process parameter to energy consumption. This information provides invaluable insights into the levers that can be adjusted to reduce the energy consumption of ACC processes. In essence, this study offers a transformative approach to enhancing the energy efficiency of ACC plants. By leveraging AI, it provides a means to predict and optimize energy consumption, offering a pathway to make ACC more cost-effective and environmentally sustainable. The findings not only facilitate the energy-efficient design of ACC plants but also prioritize parameters for adjustment, ultimately contributing to the broader goal of reducing carbon emissions and combating climate change.

Hussin et al. embarked on a crucial exploration of a rapidly evolving field—the application of machine learning in carbon capture technology [28]. At the heart of this inquiry lies the global challenge of climate change and the imperative to address it. As climate change and global warming continue to be of paramount concern, swift and effective measures are needed to safeguard our planet for future generations.

Carbon capture technology represents one of the essential tools in the fight against climate change, as it promises to reduce CO_2 emissions and mitigate the most severe effects of global warming. To achieve this, scientists, industrial sectors, and policy-makers have engaged in relentless efforts to develop innovative solutions. These endeavours involve complex processes that generate substantial amounts of data, necessitating efficient methods to analyze and predict outcomes. Machine learning, with its mathematical and statistical approach, plays a pivotal role in addressing these challenges. It offers rapid results for big data prediction and cost-efficient tools to optimize research and development.

In their study, Hussin and his team conducted a systematic review and bibliometric analysis to comprehensively examine the research landscape. They delved into crucial aspects, such as keywords, the number of publications, citations, countries, and authorship. This information is instrumental for guiding future research directions in this critical area. The key findings from this analysis highlight the leading countries in the field, including the United States, China, Iran, Canada, and the United Kingdom, each making significant contributions with the highest number of publications and citations. Moreover, the study identifies the most prevalent keywords in the selected articles, with the top six keywords being machine learning, artificial neural network, CO_2 capture, CO_2

solubility, metal-organic frameworks (MOFs), and carbon capture and storage.

The insights gained from this comprehensive study offer valuable guidance to research communities. They provide an overview of global research trends, highlight current innovations in technology, and pinpoint active research areas and hot topics. This information serves as a compass, directing researchers toward impactful areas where smart, advanced technology, particularly machine learning, can make a substantial contribution to the fight against climate change. By harnessing these innovative tools, we move closer to achieving the goal of net-zero emissions and a sustainable future for our planet.

Chen et al. embarked on a journey that underscores the immense potential of artificial intelligence (AI) in shaping the future of clean energy and carbon capture, utilization, and storage (CCUS) [29]. In an era where zero-carbon energy and negative emission technologies are pivotal to achieving a carbon-neutral future, the role of nanomaterials takes centre stage. Nanomaterials have proven to be catalysts for advancements in these transformative technologies. The critical juncture arrives with the explosive growth in data, creating an ideal landscape for the adoption and exploitation of AI within the materials research framework. The infusion of AI brings about a profound impact, revolutionizing the paradigms governing material discovery. AI empowers scientists to significantly accelerate every phase of material development and explore the vast design space that was once daunting.

This comprehensive review encapsulates the recent strides made in applying AI to the realm of nanomaterials. It delves into the specifics, emphasizing select applications of AI and nanotechnology in realizing a net-zero emission future. The selected applications span a wide spectrum, ranging from the development of solar cells and hydrogen energy to battery materials for renewable energy and CO_2 capture and conversion materials for CCUS technologies. While commending the advancements, the study candidly addresses the limitations and challenges that persist in the current landscape of AI applications in this field. It identifies the gaps that demand attention and innovation, ensuring that AI can fulfil its transformative potential.

As a guiding light for future research, the study outlines the prospects for advancements in nanomaterials. It envisions the large-scale applications of artificial intelligence, heralding a future where AI and nanotechnology work in tandem to accelerate the transition to clean energy and pave the way for effective carbon capture, utilization, and storage. This synthesis of AI and nanomaterials is not merely a scientific pursuit; it is a beacon of hope for a sustainable and carbon-neutral world.

Davies et al. presented a technical review that sheds light on the intersection of process modelling and machine learning in the realm of hydrogen production with carbon capture, specifically focusing on blue hydrogen [30]. This study provides a meticulous examination of the current landscape and the envisioned role of machine learning in research and development within the blue hydrogen production sector. Blue hydrogen, with its promise of low carbon emissions, stands as a beacon in the transition towards sustainable energy. The study aims to accurately convey the pivotal role that machine learning is poised to play in shaping the future of blue hydrogen production.

The paper carefully unveils the multifaceted application of machine learning at both material and process development levels. It begins by offering a concise overview of the prevailing trends in blue hydrogen production, introducing the fundamentals of machine learning and process modelling within this context. It then delves into the core "tools" of machine learning and process modelling, providing a foundation for the subsequent in-depth exploration of their implementation in blue hydrogen production. This comprehensive analysis highlights the merits and demerits of incorporating these techniques into the blue hydrogen production process.

Ultimately, the paper presents a clear picture of the advancements in machine learning and the pivotal role it is poised to play in expediting research and development in blue hydrogen production, spanning both material and process development fronts. As the world seeks cleaner and more sustainable energy solutions, machine learning emerges as a potent tool, offering key advantages that can propel blue hydrogen research to new heights. This study serves as a guiding example, illuminating the path towards a future where blue hydrogen production is enriched by the contributions of machine learning and where cleaner and more efficient energy sources become a reality.

Yang et al. presented a critical review that shines a spotlight on the invaluable role of machine learning in the quest for effective carbon dioxide (CO_2) capture [31]. As we strive to close the global carbon cycle, promote sustainable energy production, and meet the ambitious 1.5°C climate goal by 2050, CO_2 capture stands as a critical technology. However, the diversity and complexity of materials used in this endeavour demand innovative solutions, and machine learning has emerged as a powerful tool to address these challenges.

This critical review begins by providing an essential context, laying out the technical foundations for the application of machine learning in the field of CO_2 capture. It underscores the significant impact of machine learning-based methods on the screening and design of materials, which are instrumental in achieving efficient CO_2 capture. The review takes a comprehensive approach by categorizing materials into two major groups: adsorbents, which encompass a range of materials including metal organic frameworks, carbonaceous materials, polymers, and zeolites, and absorbents, which involve ionic liquids, amine-based absorbents, and deep eutectic solvents. The authors meticulously examine the applications of machine learning within each of these domains, offering a thorough analysis of how this effective tool is being utilized in the development of CO_2 capture techniques.

However, the review does not merely extol the virtues of machine learning but also highlights key challenges that need to be addressed for further progress. These challenges include the need for consistent and integrated databases, the intelligent digitalization of material properties, and the rigorous validation of machine learning-derived results under practical, real-world scenarios. In essence, this critical review serves as a bridge connecting the achievements of the past with the future developments in machine learning-assisted design for CO_2 capture techniques. By addressing both the promise and the challenges of machine learning in this field, it paves the way for a more sustainable and environmentally friendly future, where effective CO_2 capture is a cornerstone in the fight against climate change.

Mohan et al. embarked on a vital endeavour to address the escalating issue of carbon dioxide (CO_2) emissions from fossil fuel combustion, which significantly contributes to global warming and climate change. Carbon capture and sequestration offer a promising solution to mitigate these greenhouse gas emissions [32]. Among the various environmentally friendly solvents, Deep Eutectic Solvents (DESs) have stood out for their potential in CO_2 capture. To build a solid theoretical foundation for understanding DES activity, accurate modelling and solubility predictions are essential in anticipating system behaviour.

In their research, they employed a sophisticated approach by combining the COSMO-RS model with machine learning techniques to predict the solubility of CO_2 in diverse deep eutectic solvents. They curated a comprehensive dataset consisting of 1973 CO_2 solubility data points across 132 different DESs, spanning various temperatures, pressures, and DES molar ratios. This dataset serves as a foundation for the refinement and development of the COSMO-RS model.

Notably, the calculated CO_2 solubility in DESs using the COSMO-RS model initially exhibits significant deviations from experimental values, with an average absolute relative deviation (AARD) of 23.4%. However, the researchers demonstrated their commitment to enhancing accuracy by introducing a multilinear regression model. This model incorporates the COSMO-RS predicted solubility and a temperature-pressure dependent parameter, which effectively reduces the AARD to 12%. Nevertheless, the true breakthrough in this study emerges from the application of a machine learning model, harnessing COSMO-RS-derived features through an artificial neural network algorithm. This novel approach delivers impressive results, achieving a remarkably low AARD of only 2.72%.

The implications of this work are profound, as it introduces a powerful tool for the design and selection of DESs in the realm of CO_2 capture and utilization. The machine learning model provides a practical means of achieving accurate predictions, offering researchers and engineers a valuable resource to expedite the development of environmentally sustainable solutions for CO_2 capture. By bridging the gap between theory and experimentation, this research takes us one step closer to realizing effective carbon capture methods that can make a substantial impact on the path to a greener future.

Tarek et al. embarked on a promising journey in the realm of carbon capture by focusing on deep eutectic solvents (DESs), an environmentally friendly class of solvents. Given the critical role of DESs in various practical applications for carbon dioxide (CO_2) capture, understanding their CO_2 solubility is of paramount importance [33].

In their study, they employed a machine learning approach based on a multilayer perceptron (MLP) using molecular descriptors derived from the Conductor-like Screening Model for Real Solvents (COSMO-RS) to predict the CO_2 solubility of DESs. This endeavour results in the creation of an extensive database, comprising a remarkable 2327 data points collected from 94 unique DES mixtures. These DES mixtures are carefully composed from a combination of 2 anions, 17 cations, and 39 Hydrogen Bond Donors (HBDs), across 150 different compositions and operating conditions of temperatures and pressures.

This diligent research endeavour incorporates thorough statistical testing and hyperparameter tuning, culminating in an optimal MLP architecture that can predict CO_2 solubility with an impressive R2 value of 0.986 ± 0.002. The model also maintains an Average Absolute Relative Deviation (AARD) of 4.504 ± 0.507. To facilitate the accessibility and practicality of their findings, the researchers have thoughtfully integrated the MLP into an Excel spreadsheet included in the supporting information, making it user-friendly for fellow researchers and practitioners.

However, the significance of this work extends beyond predictive accuracy. This model is also employed to enable high-throughput screening of 1320 DES combinations, offering insights into the molecular design of DESs that can achieve high CO_2 solubilities. By doing so, this research paves the way for the creation of robust and precise models for predicting CO_2 solubility in novel DESs. Such models hold the potential to significantly reduce the need for costly and time-consuming experimental work, ultimately expediting the development of environmentally sustainable solutions for carbon capture. This study, with its fusion of machine learning and chemistry, represents a promising step forward in our ongoing efforts to address the challenges of CO_2 capture efficiently and effectively.

Eslam et al. delved into the fascinating realm of carbon capture, a critical technology for mitigating global warming [34]. Among various methods, the adsorption-based approach is gaining prominence. These researchers are dedicated to creating novel adsorbents with specific attributes, such as high capacity and selectivity, to expedite the design and deployment of adsorption-based CO_2 capture processes.

Their study holds a commendable objective: to identify a set of optimal adsorbents with the sought-after characteristics. In doing so, they utilized previously published data on various adsorbents. The ultimate aim is to provide a clear and interpretable guide for different stakeholders, including experts in process design, who are involved in carbon capture at different stages—simulation, lab-scale, and pilot-scale. The approach involves the collection of data from diverse sources, including research papers, literature reviews, and publicly available datasets. What makes this research truly noteworthy is their utilization of an ensemble of machine learning techniques that are not just accurate but also interpretable, addressing the common concern of 'black-box' predictions. By achieving a classification accuracy above 98%, the team generates human-interpretable patterns and rules.

These rules offer actionable insights, such as the recommendation to design chemically modified adsorbents with specific characteristics—high surface area, large pore volume, and small pore diameter for improved selectivity. Moreover, it advises on preferred operating conditions, indicating that low temperature and pressure conditions, near atmospheric, are ideal for achieving desired selectivity, while higher pressures are beneficial for capacity. In a world where we need efficient, sustainable solutions to combat climate change, the work by Eslam and their team is not just a significant step forward; it is a beacon of hope. By providing guidance for the selection of optimal adsorbents in a transparent and comprehensible manner, this research empowers researchers and stakeholders to make informed choices and accelerate the development of effective carbon capture technologies. This study is a testament to the power of machine learning and its potential to advance sustainability.

Gu et al. embarked on a fascinating journey to tackle one of the most pressing challenges of our time—carbon dioxide (CO_2) capture and separation [35]. In the quest to reduce global warming and harness the potential of CO_2 in various applications, the selective adsorption and separation of CO_2 from molecules like nitrogen (N_2) and water

(H_2O) is a formidable task due to their closely aligned kinetic diameters. Zeolites have long been recognized for their gas separation capabilities, but the introduction of metals into these zeolite frameworks, creating metal-zeolites, presents an almost infinite chemical space to explore. This diversity, stemming from different metal active sites, topologies, and Si/Al ratios, complicates the search for the optimal material to selectively capture CO_2.

In their study, the team harnessed the power of machine learning (ML) to predict the selective adsorption of $CO_2/N_2/H_2O$ on metal-zeolites, guided by the modulation of electrostatic polarization interactions. They began by assessing the stability of 208 metal-zeolites across five distinct topological structures, estimating the potential accessibility in experimental settings. Particularly, Sc-, Y-, and Zr-zeolites exhibit promise in this regard. The adsorption of CO_2 on metal sites is intriguing, with two possible configurations: linear vs. bent CO_2, dependent on the specific embedded metals. Their ML models employ descriptors associated with the zeolites, adsorbates, and metals to predict the adsorption strength index, demonstrating strong performance. These predictions align with experimental findings and systems, including Co-, Zn-, Cu-, Fe-SSZ-13.

One noteworthy outcome of their research is the identification of medium-pore-sized zeolites, typically with a pore limiting diameter of around 7 Å, anchored with metals like Zr, Nb, or Mo as promising materials for CO_2 adsorption and separation. Beyond the specific findings, this work introduces a machine learning scheme that can be applied to swiftly predict CO_2 adsorption and separation capabilities in various porous materials, expanding its impact beyond zeolites to porous metal-organic frameworks and amorphous materials. The innovative application of machine learning to such a complex and pressing problem showcases the potential of AI-driven approaches in advancing sustainability and addressing the challenges of our time. By combining materials science and artificial intelligence, Gu and her team have paved the way for more efficient and effective CO_2 capture and separation processes, offering a glimpse of hope in the fight against climate change.

Mehtab et al. harnessed the power of machine learning to address a critical challenge in the field of carbon capture: predicting CO_2 solubility in physical solvents with a high degree of accuracy [36]. This research takes on significant importance in the context of capturing carbon dioxide from concentrated streams at high pressures, as it offers a more efficient and cost-effective alternative to traditional experimental procedures.

The study begins by establishing a comprehensive database containing physical, thermodynamic, and structural properties of physical solvents. The researchers then employed a range of machine learning models, encompassing linear, nonlinear, and ensemble techniques, and subjected them to systematic cross-validation and grid search methods. Through this rigorous process, they identified Kernel Ridge Regression (KRR) as the optimal model for accurate predictions. To maximize the efficiency of their model, the team delved into the intricate details of their data. They rank descriptors based on their complete decomposition contributions, a technique facilitated by principal component analysis. By identifying the most crucial descriptors, they refined their model, resulting in a reduced order KRR (r-KRR) model with just nine Key Descriptors (KDs). This streamlined approach, focusing on the essential characteristics of the solvents, significantly enhances prediction accuracy.

The r-KRR model achieves remarkable accuracy with minimal root-mean-square error (0.0023), mean absolute error (0.0016), and a maximum R^2 of 0.999. The rigorous statistical analysis conducted throughout the study ensures the validity and reliability of both the established database and the machine learning models. This research offers an exciting avenue for making carbon capture more efficient and accessible. By employing machine learning to predict CO_2 solubility in physical solvents, it paves the way for quicker and more cost-effective evaluation of solvents for carbon capture. In doing so, it contributes to the ongoing efforts to combat climate change by mitigating CO_2 emissions from industrial processes.

9.3 Carbon Storage or Sequestration

The rapidly escalating levels of atmospheric carbon dioxide (CO_2) have ushered in a new era of environmental urgency. Climate change, with its far-reaching consequences, has cast a formidable shadow over our planet, making the need for effective carbon capture and storage (CCS) solutions more pressing than ever before. In the quest for sustainable climate action, carbon storage and sequestration have emerged as pivotal components, holding the promise of mitigating the impacts of greenhouse gas emissions.

Carbon capture and storage is a multifaceted approach aimed at removing and safely storing CO_2 emissions, primarily from industrial processes and power generation facilities, to prevent their release into the atmosphere [37]. This technology can significantly contribute to achieving the global climate goals set forth in agreements like the Paris Agreement, which strives to limit global warming to well below 2 degrees Celsius above pre-industrial levels [38].

The storage and sequestration of CO_2 are integral parts of this transition [39]. They offer a way to bridge the gap between continued energy production and environmental protection. There are numerous ways to achieve carbon storage, encompassing geological storage in underground formations, ocean storage, and even mineralization of CO_2 [40]. Each method has its distinct advantages, challenges, and opportunities [41].

In this review, we explore the cutting-edge technologies and innovations in the realm of carbon storage and sequestration. We delve into geological and ocean-based storage solutions, investigating emerging techniques that not only secure CO_2 but also explore ways to utilize it beneficially. Moreover, we will discuss the evolving policy landscape and international efforts to bolster CCS implementation. As the world races against time to confront climate change, it is of paramount importance to assess the state of the art in CCS, recognize the potential of nascent technologies, and chart a course for a sustainable, carbon-conscious future.

The research by Chen et al. focuses on addressing the crucial aspect of monitoring in the context of geologic carbon dioxide (CO_2) sequestration, which is essential for managing the risks associated with this environmentally significant process [42]. The study underscores the importance of effective monitoring to ensure the safe and permanent storage of CO_2 during the entire lifespan of a geologic CO_2 sequestration project. The central challenges addressed in this research pertain to determining the optimal locations for placing monitoring wells and deciding what types of data should be collected to monitor the CO_2 sequestration process accurately. Importantly, the study takes into account the uncertainties inherent in geologic sequestration sites, as these uncertainties can impact monitoring strategies and outcomes significantly.

To tackle these challenges, the researchers developed a novel approach based on filtering-based data assimilation. This procedure is designed to formulate and refine effective monitoring strategies that can adapt to the evolving conditions of CO_2 sequestration sites. To make this process computationally efficient, the study incorporates a machine learning algorithm called Multivariate Adaptive Regression Splines. This machine learning technique is employed to create reduced-order models using the results obtained from full-physics numerical simulations of CO_2 injection into saline aquifers and subsequent multi-phase fluid flow. These reduced-order models serve to streamline the monitoring strategy design, making it more practical and less computationally intensive.

The research also includes practical applications of this approach by considering example scenarios of CO_2 leakage through legacy wellbores. By analyzing these scenarious, the study demonstrates how the proposed monitoring framework can be effectively applied to reduce uncertainties associated with CO_2 leakage. The authors showcase their approach through two synthetic examples, including a simple validation case and a more complex scenario involving multiple monitoring wells. These examples demonstrate that the developed approach is capable of creating monitoring strategies that not only ensure the safety and permanence of

CO_2 storage but also consider and mitigate uncertainties in the process. In summary, the study offers a promising method for designing robust and effective monitoring systems in geologic CO_2 sequestration projects, which is vital for addressing environmental concerns and advancing the practical implementation of this technology.

The research by Menad et al. centres around the critical issue of predicting the solubility of carbon dioxide (CO_2) in brine, with direct relevance to the field of carbon capture and sequestration (CCS) in saline aquifers [43]. CCS is increasingly regarded as a pivotal strategy for mitigating CO_2 emissions and reducing the atmospheric concentration of this greenhouse gas. The solubility of CO_2 in brine is a fundamental factor in effectively monitoring and understanding the success of CO_2 sequestration.

In this study, the authors undertook the task of modelling CO_2 solubility in brine, with the input parameters being the molality of NaCl (sodium chloride), pressure, and temperature. To achieve this, they employed two advanced machine learning systems: the multilayer perceptron (MLP) and the radial basis function neural network (RBFNN). Furthermore, they implemented optimization algorithms to fine-tune these models, specifically the Levenberg-Marquardt (LM) algorithm for the MLP model, and genetic algorithm (GA), Particle Swarm Optimization (PSO), and Artificial Bee Colony (ABC) for the RBFNN model.

To develop and validate their models, the researchers utilized a comprehensive experimental dataset comprising 570 data sets from existing literature. They employ a combination of graphical and statistical assessment criteria to evaluate the performance of the models they propose. The results of their study demonstrate that all the proposed techniques exhibit excellent agreement with the experimental data, indicating their capability to accurately predict the solubility of CO_2 in brine. Notably, the performance analyses reveal that the RBFNN-ABC model outperforms other intelligent approaches and well-known models in predicting CO_2 solubility in brine. This particular model achieves a root mean square error (RMSE) value of 0.0289 and an R-squared (R^2) value of 0.9967, attesting to its high accuracy. Additionally, the authors confirmed the validity of the RBFNN-ABC model and used it to identify a small number of potentially erroneous data points, enhancing the overall quality of the dataset.

In summary, this research article contributes significantly to the field of CCS by developing advanced machine learning models for predicting CO_2 solubility in brine, a critical aspect of monitoring CO_2 sequestration. The study's findings not only provide valuable tools for understanding and optimizing CCS processes but also highlight the effectiveness of the RBFNN-ABC model in this context, paving the way for improved carbon capture and sequestration practices.

Zhang et al. addressed the crucial issue of accurately characterizing the interfacial tension (IFT) between carbon dioxide (CO_2) and brine in the context of carbon capture and sequestration (CCS) in saline aquifers [44]. The IFT is a fundamental parameter that significantly impacts the storage capacity of CO_2 in saline aquifers and is therefore essential for optimizing the design of CO_2 sequestration projects.

The authors proposed an innovative solution to this challenge by utilizing Extreme Gradient Boosting (XGBoost) trees, a form of supervised machine learning, to rapidly and accurately model the CO_2-brine IFT. The results of their study showcase the efficacy of this novel model, which not only estimates the IFT but also captures the underlying relationships between the IFT and each input variable with remarkable accuracy. To assess the performance of their model, the researchers employed statistical matrices and point-wise error analyses, which consistently demonstrate that their new approach outperforms previous machine learning methods significantly.

In addition to establishing a robust model for CO_2-brine IFT, the authors extended the application of their model to determining the optimal depth for CO_2 sequestration in saline aquifers. Their findings reveal valuable insights, indicating that higher pressure and/or lower geothermal gradients lead to a notable increase in the maximum structural trapping capacity, which occurs at shallower formations. This information is critical for making informed decisions regarding the depth at which CO_2 should be sequestered to maximize storage efficiency in saline aquifers.

In summary, this research paper contributes significantly to the field of CCS by introducing an advanced machine learning approach (XGBoost) for accurately modelling the CO_2-brine IFT. The results not only provide an effective means of characterizing this essential parameter but also offer insights into optimizing the depth of CO_2 sequestration in saline aquifers. This work has the potential to enhance the design and implementation of CO_2 sequestration projects, contributing to the reduction of CO_2 emissions and environmental impact.

Wang et al. focused on the essential task of inferring carbon dioxide (CO_2) saturation levels using a combination of synthetic surface seismic and downhole monitoring data, employing machine learning techniques for the purpose of detecting potential leakage at CO_2 sequestration sites [45]. This endeavour is of utmost importance when applying seismic methods for verification and accounting purposes in the context of CO_2 sequestration. These purposes include verifying the total injected CO_2 volume, comparing these volumes with model predictions to assess concordance, tracking the migration of the CO_2 plume within the storage reservoir, and detecting any potential leakage from the storage site.

In their study, the authors worked with synthetic data generated from 6,000 numerical multi-phase flow simulations. These simulations simulate hypothetical scenarios of CO_2 and brine leakage from a legacy well into shallow aquifers at a model CO_2 storage site. The researchers performed rock physics modelling to estimate how the seismic velocity changes due to the simulated CO_2 and brine leakage at each time step within the flow simulation outputs. This results in a total of 120,000 forward seismic velocity models.

To create synthetic seismic data, 2D finite-difference acoustic wave modelling is conducted for each velocity model, producing synthetic shot gathers. These synthetic seismic data are generated along a sparse 2D seismic line with only 5 shots and 40 receivers. The next step involves extracting six time-lapse seismic attribute anomalies from each trace within the relevant time window for each geologic layer. These seismic features are then combined with downhole pore pressure and Total Dissolved Solids (TDS) features to train machine learning algorithms.

The study also investigates the impact of seismic noise on the performance of the trained machine learning models. Despite the presence of noise, the inferred CO_2 saturations from the trained classifiers demonstrate good agreement with the actual observations. Furthermore, the inclusion of direct pressure and TDS measurements from downhole monitoring enhances the accuracy of the inferred CO_2 saturation classes obtained from the forward-modelled 2D surface seismic data.

In summary, the research by Wang and his team presents a promising approach to combine data from multiple monitoring techniques, including seismic monitoring, to achieve more accurate seismic quantitative interpretation. This methodology holds great potential for enhancing the monitoring and detection of CO_2 leakage in the context of CO_2 sequestration, thereby contributing to the safety and effectiveness of carbon capture and storage efforts.

The research article by Sun et al. is focused on the modelling of carbon dioxide (CO_2) geo-sequestration and specifically investigates the estimation of contact angles in ternary systems comprising brine, CO_2, and different mineral substrates (feldspar, mica, calcite, and quartz) [46]. The study addresses a wide range of operating conditions, including variations in salinity, pressure, and temperature. To achieve this, the researchers propose and employ two algorithms: Extreme Learning Machine (ELM) and Gradient Tree Boosting (GTB).

The primary objective of these algorithms is to simulate the wettability behaviour of CO_2/brine/mineral systems, which involves estimating receding, advancing, and static contact angles on the various mineral surfaces. Wettability, specifically contact angle, plays a critical role in understanding the interaction between CO_2, brine, and minerals in geological formations, which is vital for effective CO_2 sequestration. To assess the proficiency and accuracy of the ELM and GTB models, the researchers conducted both statistical and graphical evaluations. These evaluations reveal that the ELM model achieves a high coefficient of determination (R-squared, R^2) of 0.993, indicating its superior performance in contact angle estimation compared to the GTB model, which still performs well with an R^2 of 0.988.

Furthermore, the study investigates how different operating conditions affect the accuracy of the GTB and ELM algorithms. It is found that higher pressure and temperature can influence the accuracy of the GTB algorithm in predicting contact angles, whereas the ELM algorithm maintains acceptable performance across all tested operating conditions. Additionally, the research employs sensitivity analysis to identify the most and least influential parameters in determining the contact angle between brine, CO_2, and mineral substrates. The findings indicate that the type of mineral and temperature are the most and least significant parameters, respectively, in the determination of contact angles within the brine/CO_2/mineral ternary system.

In summary, the work carried out by Sun and the team introduces advanced machine learning algorithms (ELM and GTB) to model CO_2 geo-sequestration and estimate contact angles in ternary systems involving brine, CO_2, and different minerals. These findings contribute to a better understanding of the wettability behaviour in CO_2 sequestration processes, helping to optimize the management of CO_2 in geological formations.

The research article by Yuan et al. delves into the application of machine learning for the prediction of carbon dioxide (CO_2) adsorption on a class of materials known as Biomass Waste-Derived Porous Carbons (BWDPCs) [47]. BWDPCs are versatile materials used in sustainable waste management and carbon capture processes. However, the complexity of BWDPCs, including their diverse textural properties, the presence of various functional groups, and the variations in temperature and pressure conditions during CO_2 adsorption, makes it challenging to comprehend the underlying mechanisms governing CO_2 adsorption on these materials.

To address this challenge, the researchers compiled a comprehensive dataset consisting of 527 data points collected from peer-reviewed publications. They then applied machine learning techniques to systematically analyze and map the CO_2 adsorption behaviour of BWDPCs in relation to their textural and compositional properties, as well as the specific adsorption parameters involved. The study involves the development of various tree-based machine learning models, with a particular focus on Gradient Boosting Decision Trees (GBDTs). Among the models tested, GBDTs demonstrate the best predictive performance, achieving a high coefficient of determination (R-squared, R^2) of 0.98 on the training data and 0.84 on the test data. This indicates the efficacy of GBDTs in accurately predicting CO_2 adsorption behaviour for BWDPCs.

Additionally, the researchers classified the BWDPCs in the dataset into two categories: Regular Porous Carbons (RPCs) and Heteroatom-Doped Porous Carbons (HDPCs). Once again, the GBDT model excels, with R2 values of 0.99 and 0.98 on the training data and 0.86 and 0.79 on the test data for RPCs and HDPCs, respectively. This demonstrates the versatility and accuracy of the GBDT model in predicting CO_2 adsorption behaviour for both types of BWDPCs. Moreover, the study examines the feature importance in the machine learning models, revealing the relative significance of different factors affecting CO_2 adsorption on BWDPCs. The order of precedence is determined to be adsorption parameters, textural properties, and compositional properties. This insight provides valuable guidance for the synthesis and design of porous carbons tailored for specific CO_2 adsorption applications.

In summary, the work conducted by Yuan and the research team showcases the power of machine learning in understanding and predicting CO_2 adsorption on complex materials like BWDPCs. The findings not only advance our knowledge of CO_2 adsorption mechanisms but also offer practical guidance for the development of BWDPCs optimized for carbon capture and environmental applications.

The research conducted by Thanh et al. focuses on the important task of predicting the efficiency of carbon dioxide (CO_2) storage in underground saline aquifers, with the ultimate goal of reducing atmospheric CO_2 emissions [48]. Specifically, the study concentrates on understanding the processes of CO_2 residual and solubility within deep saline aquifers, which are critical for enhancing the security and effectiveness of CO_2 storage. To achieve this objective, the researchers develop three supervised machine learning (ML)-based models: Random Forest (RF), extreme gradient boosting (XGBoost), and Support Vector Regression (SVR). These models are trained and tested using a diverse field-scale simulation database comprising 1,509 samples collected from relevant literature sources.

To evaluate and compare the prediction accuracy of the three ML models, the researchers employed graphical and statistical indicators. Based on the prediction results, the models are ranked according to their accuracy, with XGBoost demonstrating the highest performance, followed by RF and SVR. The XGBoost-based predictive model stands out by achieving an exceptionally low root mean square error (RMSE) of 0.0041 and a high correlation factor (R-squared, R^2) of 0.9993 for both residual and solubility trapping efficiency. In comparison, RF and SVR exhibit RMSEs of 0.0243 and 0.074, and R2 values of 0.9781 and 0.9284, respectively.

Furthermore, the applicability of the XGBoost model is validated, and it successfully detects only 15 suspected data points across the entire database, indicating its reliability. As a result, the proposed model can serve as a valuable template for predicting the CO_2 trapping index in various saline formations worldwide. The research goes a step further by subjecting the XGBoost model to comprehensive blind testing against reservoir simulation models. The model demonstrates its robustness and reliability in these tests, suggesting its potential utility as a screening and process planning tool for assessing the uncertainty of carbon storage projects.

In summary, Thanh and the research team leveraged knowledge-based machine learning techniques to create predictive models for CO_2 storage efficiency in underground saline aquifers. These models offer valuable insights for enhancing the security and success of CO_2 storage projects and have the potential to serve as a practical tool for assessing the uncertainty associated with carbon storage endeavours.

The research article by Artun et al. delves into the realm of carbon capture and sequestration (CCS), a critical process aimed at capturing carbon dioxide (CO_2) emissions from refineries, industrial facilities, and major point sources, such as power plants, and storing the captured CO_2 in subsurface formations [49]. CCS is recognized as a vital technology for mitigating greenhouse gas emissions, and it holds the potential to establish an industry comparable to, or even greater than, the existing oil and gas sector.

One particular focus of CCS is the utilization of subsurface formations, including unconventional oil and gas reservoirs, to store significant quantities of CO_2. However, despite the importance of these formations within the oil and gas industry, a comprehensive understanding of CO_2 sequestration in unconventional reservoirs is still evolving. The primary objective of the research presented in this paper is to leverage a vast dataset of numerical simulation results in combination with data analytics and machine learning techniques to identify the key parameters that influence CO_2 sequestration in depleted shale reservoirs. The study employs machine learning-based predictive models, including multiple linear regression, regression tree, bagging, random forest, and gradient boosting, to predict the cumulative amount of CO_2 injected into these reservoirs.

The researchers conducted variable importance analysis to identify and rank the most crucial reservoir and operational parameters affecting CO_2 sequestration performance. The results indicate that random forest stands out as the most effective predictive model among the machine learning techniques employed, while regression tree exhibits the lowest predictive ability, primarily due to issues related to overfitting. The most influential variable for predicting cumulative CO_2 sequestration is found to be stimulated reservoir volume fracture permeability, highlighting its significance in shaping the success of CO_2 storage in shale reservoirs.

In conclusion, the workflows, machine learning models, and findings presented in this study offer valuable insights for exploration and production companies interested in quantifying the performance of CO_2 sequestration in depleted shale reservoirs. The research contributes to the ongoing development of knowledge and technology in the field of carbon capture and sequestration, which is crucial for addressing global climate change and promoting sustainable energy practices.

The research conducted by Ahmed Farid Ibrahim focuses on the crucial task of understanding the wettability behaviour of carbon dioxide (CO_2) in shale formations, which has significant implications for a variety of applications, including CO_2-enhanced oil recovery, CO_2 foam hydraulic fracturing, and CO_2 storage in saline aquifers and shale formations [50]. Traditional laboratory methods for determining shale wettability, such as contact angle measurements and nuclear magnetic resonance spectroscopy, are known to be time-consuming and expensive. To overcome these limitations, the study employs machine learning (ML) techniques to estimate shale wettability based on shale characteristics and experimental conditions.

In the pursuit of this objective, the research team compiled a dataset comprising various shale samples tested under different conditions, including variations in pressure, temperature, and brine salinity. The goal is to predict shale wettability. To evaluate the relationship between the contact angle value and other input parameters, the researchers utilized Pearson's correlation coefficient and linear regression analysis. ML models, specifically decision trees, random forests, function networks, and gradient boosting regressors, are then employed for the prediction of shale wettability. The findings of the research highlight several influential factors that affect shale wettability, including operating pressure, temperature, rock mineralogy, and total organic content. While the linear regression model exhibits limited accuracy, the ML models, particularly random forests, decision trees, gradient boosting regressors, and function networks, prove to be highly effective in predicting the contact angle value, yielding high R-squared (R^2) values. Among these models, the gradient boosting regressor stands out with the best performance, achieving R^2 values of 0.99 for the training dataset and 0.98 for the testing dataset, with a root mean square error below 5 degrees. Sensitivity analysis reveals that pressure has a significant impact on shale wettability, with low pressures resulting in water-wet shale, regardless of other shale properties.

In conclusion, the study underscores the efficiency of the developed ML models in predicting shale wettability within the context of CO_2-water-shale systems. This approach offers a faster and more cost-effective alternative to traditional experimental analyses, enabling a better understanding of shale wettability for a range of applications related to CO_2 in geological formations.

The research article by Jianchun et al. introduces a novel and automated approach for the development of surrogate flow models in the context of carbon dioxide (CO_2) sequestration [51]. These surrogate models are essential for improving the efficiency and economic benefits of subsurface resource utilization by simulating multiphase flow in heterogeneous porous media. Numerical simulation, while powerful, often demands significant computational resources and time due to the inherent high-dimensional nonlinearity, heterogeneity, and the coupling of multiple physical processes.

Traditionally, creating surrogate models that accurately represent the complex behaviour of subsurface systems requires extensive human intervention and a series of trial-and-error processes, even for experts in various domains. To address this challenge, the study presents an automated surrogate flow model workflow named Surrogate Flow Model Search (SFMS). SFMS leverages deep learning techniques to automate many of the intricate and time-consuming tasks associated with model development. This includes the design of neural architectures and loss functions. By incorporating neural architectures and loss functions into a joint hyperparameter optimization process, SFMS streamlines the creation of surrogate flow models. This workflow not only accelerates the model development process but also eliminates the need for deep-learning expertize among researchers. As a result, SFMS empowers researchers to construct high-quality surrogate models more efficiently.

The study demonstrates the effectiveness of SFMS using an example involving a CO_2 injection into saline aquifers. The results illustrate that SFMS can automatically generate highly accurate surrogate flow models in a short timeframe (less than 1300 seconds). These models are capable of predicting the behaviour of CO_2 injection across 120-time steps under various well controls and placements with a low average relative error (less than 0.4%). In essence, SFMS significantly reduces the time and effort required to develop precise and reliable surrogate models, presenting a valuable approach to surrogate model development.

Moreover, the research article by Tariq et al. addresses the critical issue of effectively storing carbon dioxide (CO_2) in geological formations [52]. This storage process is particularly relevant when combined with algal-based removal technology, as it enhances carbon capture efficiency, leverages biological processes for sustainable and long-term sequestration, and contributes to ecosystem restoration. However, the success of geological carbon storage hinges on the interactions and wettability of the various components involved, including rock, CO_2, and brine. The wettability of the rock during storage determines crucial factors such as CO_2/brine distribution, maximum storage capacity, and trapping potential.

Traditionally, assessing the wettability of CO_2 on storage or caprocks experimentally is challenging due to the high reactivity of CO_2 and the associated risks of damage. To address this challenge, the study employs data-driven machine learning (ML) models as a more efficient and less hazardous alternative. These ML models enable research at geological storage conditions that are either impossible or dangerous to achieve in a laboratory setting. The research employs robust ML models, including Fully Connected Feedforward Neural Networks (FCFNNs), extreme gradient boosting, k-nearest neighbours, decision trees, adaptive boosting, and random forest, to model the wettability of the CO_2/brine and rock minerals, specifically quartz and mica, in a ternary system under varying conditions. The study includes exploratory data analysis methods to examine the experimental data and implements techniques such as GridSearchCV and Kfold cross-validation to enhance the performance of the ML models. Additionally, sensitivity plots are generated to understand the influence of individual parameters on the model's performance. The results of the study indicate that the applied ML models effectively predict the wettability behaviour of the mineral/CO_2/brine system under various operating conditions. Among the ML techniques, the fully connected feedforward neural network (FCFNN) stands out as the most accurate, achieving an R-squared (R^2) value above 0.98 and an error of less than 3%. This demonstrates the proficiency of the FCFNN model in accurately capturing the wettability dynamics within the CO_2/brine and mineral system, offering valuable insights for carbon geo-sequestration endeavours.

The research conducted by Surasani and the team focuses on the field of Geological Carbon Sequestration (GCS), where carbon dioxide (CO_2) is injected and stored in geological formations as a means of mitigating greenhouse gas emissions [53]. Performing experiments to study GCS can be economically and scientifically impractical due to the complex and extensive geological processes involved. Traditional numerical simulation methods have been employed in the past, but they often prove to be computationally intensive and time-consuming, particularly when dealing with intricate geometrical models.

In this study, the researchers explored a novel approach by integrating traditional numerical simulation results with machine learning techniques to predict the future trends of critical output parameters in GCS processes. The primary simulations utilize multiphase flow modelling to cover the entire geological time scale, offering insights into how CO_2 interacts with geological formations over time. To forecast the trends of output parameters during the post-injection phase, the research employs time series neural networks. These networks take CO_2 sequestration parameter values as input and target data to make predictions. Specifically, recurrent neural network models are used in this context, and various training algorithms are tested.

The results of the study demonstrate that the recurrent neural network models, particularly when using the Levenberg-Marquardt (LM) algorithm for training, provide reasonable predictions for the output variables. The accuracy of the predictions is validated using performance metrics such as Root Mean Square Error (RMSE) and R-squared values. For instance, the R2 and RMSE values for the NAR (Nonlinear Auto Regressive) model in the context of structural and residual trapping are both very high, indicating the model's capability to accurately predict these critical aspects of GCS. The R^2 values for structural and residual trapping are 0.9801 and 0.9805, with RMSE values of 0.0515 and 0.0506, respectively.

This research represents an initial exploration of the integration of machine learning techniques into the analysis of GCS for heterogeneous reservoir models. It showcases the potential of machine learning in enhancing our understanding of complex geological processes, particularly those related to carbon sequestration in geological formations.

Yamada et al. addressed the challenge of Carbon Capture and Storage (CCS) in depleted gas reservoirs, which are considered attractive sites for CO_2 storage due to their substantial capacity, proven seal integrity, existing infrastructure, and availability of subsurface data [54]. However, injecting CO_2 into depleted formations may potentially lead to the formation of hydrates near the wellbore, primarily due to the Joule-Thomson cooling effect, which could create injectivity issues.

The study's primary objective is to propose a novel approach for assessing the risk of hydrate formation during CO_2 injection into depleted gas reservoirs using a physics-based Machine Learning (ML) approach. The researchers began by selecting input parameters for the ML models based on sensitivity study results using an analytical solution, considering various operational and petrophysical values. Subsequently, the ML models are fine-tuned and tested using datasets derived from numerical reservoir simulation results that cover a wide range of input parameter values. A notable aspect of this research is that it represents the first known instance of applying ML techniques for risk assessment related to CO_2 hydrate formation in depleted gas reservoirs. The ML models developed in this study demonstrate efficient performance in predicting hydrate-forming events. Among the models, the deep neural network model stands out as the most effective, achieving a recall value of 95% and a precision value of 84%. These results highlight the potential of ML models for risk assessment during the screening stage of CCS projects.

The research suggests a promising workflow for future CCS endeavours, wherein ML-based screening can be combined with detailed analysis through numerical simulations, particularly for high-risk cases. This approach offers an efficient means of probing and assessing the risks associated with hydrate formation during CO_2 injection into depleted gas reservoirs, contributing to the advancement of CCS technologies and the sustainable storage of CO_2 in geological formations.

9.3.1 Advantages and Disadvantages

- Advantages of ML and AI in CCS:
    1. **Enhanced Efficiency:** ML and AI can significantly improve the efficiency of CCS operations. They excel in optimizing complex processes, such as reservoir monitoring, by continuously analyzing data, making predictions, and automating decision-making. This enhanced efficiency results in cost savings and better resource utilization.
    2. **Improved Prediction and Risk Assessment:** ML models are capable of making accurate predictions and identifying potential issues in CCS projects. For example, they can foresee events like CO2 leakage, hydrate formation, or pressure imbalances. This proactive approach enables operators to address these challenges before they become critical, minimizing risks.
    3. **Real-time Monitoring:** ML and AI systems enable real-time monitoring of CCS activities. They continuously analyze data streams and sensor readings, allowing for the early detection of anomalies or deviations from expected performance. Timely alerts and automated responses can prevent accidents or operational disruptions.
    4. **Data Analysis:** CCS projects generate vast amounts of data, including geological and operational data. ML algorithms excel at analyzing this data to extract valuable insights and patterns. They help in reservoir characterization, fault detection, and predictive maintenance, contributing to effective reservoir management.
- Disadvantages of ML and AI in CCS:
    1. **Data Dependency:** The accuracy of ML and AI models is highly dependent on the availability and quality of data. In some CCS projects, obtaining sufficient historical and real-time data for training can be a challenge. Limited data can hinder the development of accurate models.
    2. **Model Complexity:** Developing and maintaining ML models can be resource-intensive. It often requires specialized expertize

in data science and machine learning. Complex models might be challenging to implement, particularly for organizations with limited technical resources.

3. **Uncertainty:** While ML models are powerful tools, they are not infallible. They provide predictions with a degree of uncertainty, and there can be variations in model accuracy. This introduces an element of uncertainty into CCS operations and decision-making processes.
4. **Model Validation:** Ensuring the accuracy and reliability of ML models in real-world CCS applications can be complex. Continuous model validation and adaptation are necessary to account for changes in geological conditions, operational parameters, and other dynamic factors.

In summary, ML and AI offer tremendous potential for enhancing CCS processes by increasing efficiency, predicting risks, and enabling real-time monitoring. However, the challenges of data availability, model complexity, uncertainty, and validation must be addressed to fully realize the benefits of these technologies in the context of CCS projects. Effective implementation and ongoing model refinement are critical for successful integration.

9.4 Conclusion

This chapter explores the burgeoning field of applying ML and AI to CCS, delving into the myriad ways in which these technologies are revolutionizing the landscape of carbon capture and storage. We will elucidate their application in improving the design and operation of CCS facilities, ensuring secure and leak-free storage, and optimizing the overall CCS lifecycle [5]. Through a comprehensive review of existing studies and a critical assessment of the state of the art, we aim to shed light on the immense possibilities and ongoing challenges of marrying ML and AI with CCS technology. To navigate this fascinating realm, we will embark on a journey that explores the data-driven insights and predictive models that can significantly enhance CCS systems.

References

[1] Wei-Yin Chen, John Seiner, Toshio Suzuki and Maximilian Lackner. 2012. Handbook of Climate Change Mitigation. https://doi.org/https://doi.org/10.1007/978-1-4419-7991-9.

[2] Zhang, Z., Wang, T., Blunt, M. J., Anthony, E. J., Park, A. H. A., Hughes R. W. et al. 2020. Advances in carbon capture, utilization and storage. Appl. Energy 278. https://doi.org/10.1016/j.apenergy.2020.115627.

[3] Shaik, N. B., Sayani, J. K. S., Benjapolakul, W., Asdornwised, W., Chaitusaney, S. 2022. Experimental investigation and ANN modelling on CO_2 hydrate kinetics in multiphase pipeline systems. Sci. Rep. 12. https://doi.org/10.1038/s41598-022-17871-z.

[4] Sayani, J. K. S., Lal, B. and Pedapati, S. R. 2022. Comprehensive review on various gas hydrate modelling techniques: prospects and challenges. Archives of Computational Methods in Engineering 29: 2171–2207. https://doi.org/10.1007/s11831-021-09651-1.

[5] Krishnan, A., Nighojkar, A. and Kandasubramanian, B. 2023. Emerging towards zero carbon footprint via carbon dioxide capturing and sequestration. Carbon Capture Science and Technology 9. https://doi.org/10.1016/j.ccst.2023.100137.

[6] Osman, A. I., Hefny, M., Abdel Maksoud, M. I. A., Elgarahy, A. M. and Rooney, D. W. 2021. Recent advances in carbon capture storage and utilisation technologies: a review. Environ Chem. Lett. 19: 797–849. https://doi.org/10.1007/s10311-020-01133-3.

[7] Du, K., Xie, C. and Ouyang, X. 2017. A comparison of carbon dioxide (CO_2) emission trends among provinces in China. Renewable and Sustainable Energy Reviews 73: 19–25. https://doi.org/10.1016/j.rser.2017.01.102.

[8] Yao, P., Yu, Z., Zhang, Y. and Xu, T. 2023. Application of machine learning in carbon capture and storage: An in-depth insight from the perspective of geoscience. Fuel 333. https://doi.org/10.1016/j.fuel.2022.126296.

[9] Bhavsar, A., Hingar, D., Ostwal, S., Thakkar, I., Jadeja, S. and Shah, M. 2023. The current scope and stand of carbon capture storage and utilization ~ A comprehensive review. Case Studies in Chemical and Environmental Engineering 8. https://doi.org/10.1016/j.cscee.2023.100368.

[10] Paraschiv, S. and Paraschiv, L. S. 2020. Trends of carbon dioxide (CO_2) emissions from fossil fuels combustion (coal, gas and oil) in the EU member states from 1960 to 2018. Energy Reports 6: 237–242. https://doi.org/10.1016/j.egyr.2020.11.116.

[11] Wei, K., Guan, H., Luo, Q., He, J. and Sun, S. 2022. Recent advances in CO_2 capture and reduction. Nanoscale 14: 11869–11891. https://doi.org/10.1039/d2nr02894h.

[12] Bert Metz, Ogunlade Davidson, Heleen de Coninck and Manuela Loos. n.d. IPCC Special Report on Carbon Dioxide Capture and Storage.

[13] Mirparizi, M., Shakeriaski, F., Salehi, F. and Zhang, C. 2023. Available challenges and recent progress in carbon dioxide capture, and reusing methods toward renewable energy. Sustainable Energy Technologies and Assessments 58. https://doi.org/10.1016/j.seta.2023.103365.

[14] Anderson, R., Rodgers, J., Argueta, E., Biong, A. and Gómez-Gualdrón, D. A. 2018. Role of pore chemistry and topology in the CO_2 capture capabilities of MOFs: From molecular simulation to machine learning. Chemistry of Materials 30: 6325–6337. https://doi.org/10.1021/acs.chemmater.8b02257.

[15] Dureckova, H., Krykunov, M., Aghaji, M. Z. and Woo, T. K. 2019. Robust machine learning models for predicting high CO_2 working capacity and CO_2/H_2 selectivity of gas adsorption in metal organic frameworks for precombustion carbon capture. Journal of Physical Chemistry C 123: 4133–4139. https://doi.org/10.1021/acs.jpcc.8b10644.

[16] Leperi, K. T., Yancy-Caballero, D., Snurr, R. Q. and You, F. 2019. 110th Anniversary: surrogate models based on artificial neural networks to simulate and optimize pressure swing adsorption cycles for CO_2 capture. Ind. Eng. Chem. Res. 58: 18241–18252. https://doi.org/10.1021/acs.iecr.9b02383.

[17] Rahimi, M., Moosavi, S. M., Smit, B. and Hatton, T. A. 2021. Toward smart carbon capture with machine learning. Cell Rep. Phys. Sci. 2. https://doi.org/10.1016/j.xcrp.2021.100396.

[18] Helei, L., Tantikhajorngosol, P., Chan, C. and Tontiwachwuthikul, P. 2021. Technology development and applications of artificial intelligence for post-combustion carbon dioxide capture: Critical literature review and perspectives. International Journal of Greenhouse Gas Control 108. https://doi.org/10.1016/j.ijggc.2021.103307.

[19] Yan, Y., Borhani, T. N., Subraveti, S. G., Pai, K. N., Prasad, V., Rajendran, A. et al. 2021. Harnessing the power of machine learning for carbon capture, utilisation, and storage

(CCUS)-a state-of-the-art review. Energy Environ. Sci. 14: 6122–6157. https://doi.org/10.1039/d1ee02395k.

[20] Fathalian, F., Aarabi, S., Ghaemi, A. and Hemmati, A. 2022. Intelligent prediction models based on machine learning for CO_2 capture performance by graphene oxide-based adsorbents. Sci. Rep. 12. https://doi.org/10.1038/s41598-022-26138-6.

[21] Irfan, S. A., Abdulkareem, F. A., Radman, A., Faugere, G. and Padmanabhan, E. 2022. Artificial neural network (ANN) modeling for CO_2 adsorption on Marcellus Shale. *In*: IOP Conf Ser Earth Environ Sci, Institute of Physics. https://doi.org/10.1088/1755-1315/1003/1/012029.

[22] Gupta, S. and Li, L. 2022. The potential of machine learning for enhancing CO_2 sequestration, storage, transportation, and utilization-based processes: a brief perspective. JOM 74: 414–428. https://doi.org/10.1007/s11837-021-05079-x.

[23] Narimani, S. 2022. Development of machine learning model for CO_2 capture plants to predict solvent degradation. University of South-Eastern Norway. www.usn.no.

[24] Priya, A. K., Devarajan, B., Alagumalai, A. and Song, H. 2023. Artificial intelligence enabled carbon capture: A review. Science of the Total Environment 886. https://doi.org/10.1016/j.scitotenv.2023.163913.

[25] Hosseinpour, M., Shojaei, M. J., Salimi, M. and Amidpour, M. 2023. Machine learning in absorption-based post-combustion carbon capture systems: A state-of-the-art review. Fuel 353. https://doi.org/10.1016/j.fuel.2023.129265.

[26] Zhang, H., Thanh, H. V., Rahimi, M., Al-Mudhafar, W. J., Tangparitkul, S., Zhang, T. et al. 2023. Improving predictions of shale wettability using advanced machine learning techniques and nature-inspired methods: Implications for carbon capture utilization and storage. Science of the Total Environment 877. https://doi.org/10.1016/j.scitotenv.2023.162944.

[27] Aliyon, K., Rajaee, F. and Ritvanen, J. 2023. Use of artificial intelligence in reducing energy costs of a post-combustion carbon capture plant. Energy 278. https://doi.org/10.1016/j.energy.2023.127834.

[28] Hussin, F., Md Rahim, S. A. N., Hatta, N. S. M., Aroua, M. K., Mazari, S. A. 2023. A systematic review of machine learning approaches in carbon capture applications. Journal of CO_2 Utilization 71. https://doi.org/10.1016/j.jcou.2023.102474.

[29] Chen, H., Zheng, Y., Li, J., Li, L. and Wang, X. 2023. AI for nanomaterials development in clean energy and carbon capture, utilization and storage (CCUS). ACS Nano 17: 9763–9792. https://doi.org/10.1021/acsnano.3c01062.

[30] George Davies, W., Babamohammadi, S., Yang, Y. and Masoudi Soltani, S. 2023. The rise of the machines: A state-of-the-art technical review on process modelling and machine learning within hydrogen production with carbon capture. Gas Science and Engineering 118. https://doi.org/10.1016/j.jgsce.2023.205104.

[31] Yang, Z., Chen, B., Chen, H. and Li, H. 2023. A critical review on machine-learning-assisted screening and design of effective sorbents for carbon dioxide (CO_2) capture, Front Energy Res. 10. https://doi.org/10.3389/fenrg.2022.1043064.

[32] Mohan, M., Demerdash, O., Simmons, B. A., Smith, J. C., Kidder, M. K. and Singh, S. 2023. Accurate prediction of carbon dioxide capture by deep eutectic solvents using quantum chemistry and a neural network. Green Chemistry 25: 3475–3492. https://doi.org/10.1039/d2gc04425k.

[33] Lemaoui, T., Boublia, A., Lemaoui, S., Darwish, A. S., Ernst, B., Alam, M. et al. 2023. Predicting the CO_2 capture capability of deep eutectic solvents and screening over 1000 of their combinations using machine learning. ACS Sustain Chem. Eng. 11: 9564–9580. https://doi.org/10.1021/acssuschemeng.3c00415.

[34] Al-Sakkari, E. G., Ragab, A., So, T. M. Y., Shokrollahi, M., Dagdougui, H., Navarri, P. et al. 2023. Machine learning-assisted selection of adsorption-based carbon dioxide capture materials. J. Environ. Chem. Eng. 11: 110732. https://doi.org/10.1016/j.jece.2023.110732.
[35] Gu, Y.-T., Gu, Y.-M., Tao, Q., Wang, X., Zhu, Q. and Ma, J. 2023. Machine learning for prediction of $CO_2/N_2/H_2O$ selective adsorption and separation in metal-zeolites. Journal of Materials Informatics 3. https://doi.org/10.20517/jmi.2023.25.
[36] Mehtab, V., Alam, S., Povari, S., Nakka, L., Soujanya, Y. and Chenna, S. 2023. Reduced order machine learning models for accurate prediction of CO_2 capture in physical solvents. Environ Sci. Technol. https://doi.org/10.1021/acs.est.3c00372.
[37] Gabrielli, P., Gazzani, M. and Mazzotti, M. 2020. The role of carbon capture and utilization, carbon capture and storage, and biomass to enable a net-zero-CO_2 emissions chemical industry. Ind. Eng. Chem. Res. 59: 7033–7045. https://doi.org/10.1021/acs.iecr.9b06579.
[38] Bataille, C., Åhman, M., Neuhoff, K., Nilsson, L. J., Fischedick, M., Lechtenböhmer, S. et al. 2018. A review of technology and policy deep decarbonization pathway options for making energy-intensive industry production consistent with the Paris Agreement. J. Clean Prod. 187: 960–973. https://doi.org/10.1016/j.jclepro.2018.03.107.
[39] Witkowski, A., Majkut, M. and Rulik, S. 2014. Analysis of pipeline transportation systems for carbon dioxide sequestration. Archives of Thermodynamics 35: 117–140. https://doi.org/10.2478/aoter-2014-0008.
[40] Li, N., Feng, W., Yu, J., Chen, F., Zhang, Q., Zhu, S. et al. 2023. Recent advances in geological storage: trapping mechanisms, storage sites, projects, and application of machine learning. Energy and Fuels 37: 10087–10111. https://doi.org/10.1021/acs.energyfuels.3c01433.
[41] Yu, X., Catanescu, C. O., Bird, R. E., Satagopan, S., Baum, Z. J., Lotti Diaz, L. M. et al. 2023. Trends in research and development for CO_2 capture and sequestration. ACS Omega 8: 11643–11664. https://doi.org/10.1021/acsomega.2c05070.
[42] Chen, B., Harp, D. R., Lin, Y., Keating, E. H. and Pawar, R. J. 2018. Geologic CO_2 sequestration monitoring design: A machine learning and uncertainty quantification based approach. Appl Energy 225: 332–345. https://doi.org/10.1016/j.apenergy.2018.05.044.
[43] Menad, N. A., Hemmati-Sarapardeh, A., Varamesh, A. and Shamshirband, S. 2019. Predicting solubility of CO_2 in brine by advanced machine learning systems: Application to carbon capture and sequestration. Journal of CO_2 Utilization 33: 83–95. https://doi.org/10.1016/j.jcou.2019.05.009.
[44] Zhang, J., Feng, Q., Zhang, X., Shu, C., Wang, S. and Wu, K. 2020. A supervised learning approach for accurate modeling of CO_2-brine interfacial tension with application in identifying the optimum sequestration depth in saline aquifers. Energy and Fuels 34: 7353–7362. https://doi.org/10.1021/acs.energyfuels.0c00846.
[45] Wang, Z., Dilmore, R. M. and Harbert, W. 2020. Inferring CO_2 saturation from synthetic surface seismic and downhole monitoring data using machine learning for leakage detection at CO_2 sequestration sites. International Journal of Greenhouse Gas Control 100. https://doi.org/10.1016/j.ijggc.2020.103115.
[46] Sun, X., Bi, Y., Guo, Y., Ghadiri, M. and Mohammadinia, S. 2021. CO_2 geo-sequestration modeling study for contact angle estimation in ternary systems of brine, CO_2, and mineral. J. Clean Prod. 283. https://doi.org/10.1016/j.jclepro.2020.124662.
[47] Yuan, X., Suvarna, M., Low, S., Dissanayake, P. D., Lee, K. B., Li, J. et al. 2021. Applied machine learning for prediction of CO_2 adsorption on biomass waste-derived porous carbons. Environ. Sci. Technol. 55: 11925–11936. https://doi.org/10.1021/acs.est.1c01849.

[48] Vo Thanh, H., Yasin, Q., Al-Mudhafar, W. J. and Lee, K. K. 2022. Knowledge-based machine learning techniques for accurate prediction of CO_2 storage performance in underground saline aquifers. Appl. Energy 314. https://doi.org/10.1016/j.apenergy.2022.118985.

[49] Hassan Baabbad, H. K., Artun, E. and Kulga, B. 2022. Understanding the controlling factors for CO_2 sequestration in depleted shale reservoirs using data analytics and machine learning. ACS Omega 7: 20845–20859. https://doi.org/10.1021/acsomega.2c01445.

[50] Ibrahim, A. F. 2023. Prediction of shale wettability using different machine learning techniques for the application of CO_2 sequestration. Int. J. Coal Geol. 276. https://doi.org/10.1016/j.coal.2023.104318.

[51] Xu, J., Fu, Q. and Li, H. 2023. A novel deep learning-based automatic search workflow for CO_2 sequestration surrogate flow models. Fuel 354. https://doi.org/10.1016/j.fuel.2023.129353.

[52] Tariq, Z., Ali, M., Hassanpouryouzband, A., Yan, B., Sun, S. and Hoteit, H. 2023. Predicting wettability of mineral/CO_2/brine systems via data-driven machine learning modeling: Implications for carbon geo-sequestration. Chemosphere 345. https://doi.org/10.1016/j.chemosphere.2023.140469.

[53] Punnam, P. R., Dutta, A., Krishnamurthy, B. and Surasani, V. K. 2023. Study on utilization of machine learning techniques for geological CO_2 sequestration simulations. Mater Today Proc. 72: 378–385. https://doi.org/10.1016/j.matpr.2022.08.109.

[54] Yamada, K., Fernandes, B. R. B., Kalamkar, A., Jeon, J., Delshad, M., Farajzadeh, R. et al. 2024. Development of a hydrate risk assessment tool based on machine learning for CO_2 storage in depleted gas reservoirs. Fuel 357. https://doi.org/10.1016/j.fuel.2023.129670.

CHAPTER 10

Decarbonizing Global Industry

Technologies and Policies

Baldeep Singh,[1,*] *Anipeddi Manjusha*[2] and *Bhajan Lal*[2]

10.1 Introduction

The urgent need to decarbonize the global industrial sector has become increasingly apparent in the battle against climate change. Industries are major contributors to greenhouse gas emissions, making it absolutely essential to transform them for the sake of a sustainable future. Achieving decarbonization is a complex process that involves several key elements: transitioning to renewable energy sources, improving energy efficiency, adopting carbon capture and storage technologies, and electrifying industrial operations. Governments, industry, and society must work closely together on this project, which must also be supported by international cooperation, clean technology investments, and supportive regulations. By adopting these tactics, industries may drastically lower their carbon emissions, address the serious issues posed by climate change, and chart a road for a more sustainable and prosperous future. Decarbonizing the global industrial sector also strengthens resilience in the face of the uncertainties posed by a carbon-constrained future by providing not only environmental advantages but also significant economic prospects. Companies that pioneer and invest in sustainable practices are poised to secure a competitive advantage, effectively adapt to evolving regulations

[1] Petroleum Engineering Department, Rajiv Gandhi Institute of Petroleum Technology, India.
[2] Chemical Engineering Department, Center for Carbon Capture, Storage and Utilization (CCUS), Institute of Sustainable Energy, Universiti Teknologi PETRONAS, Bandar Seri Iskandar, Perak, Malaysia.
* Corresponding author: sbaldeep752@gmail.com

and changing consumer preferences, and thrive in a rapidly transforming market. Achieving the transition toward cleaner industrial processes is a collective undertaking that necessitates a sustained dedication to research, development, and practical implementation [1, 2]. As public consciousness of climate issues continues to grow, and consumer demand for sustainable products surges, industries stand at a pivotal juncture, offering them a unique chance to align their objectives with those of a more eco-conscious and sustainable world.

Throughout the journey toward decarbonization, collaboration and innovation will emerge as the driving forces propelling industries to reduce their carbon emissions while prospering in a globally aware and ecologically responsible economic landscape. The drive to decarbonize the global industrial sector goes beyond mere emission reduction; it entails catalyzing a fundamental shift in how we approach both production and consumption. Embracing the principles of a circular economy, characterized by continuous material recycling, and minimized waste generation, holds the promise of simultaneously reducing emissions and alleviating resource depletion. Realizing this transformation necessitates the reconfiguration of products and processes, a reimagining of business models, and active engagement with consumers to promote responsible consumption patterns. Moreover, the significance of education and awareness cannot be overstated; a well-informed public has the potential to catalyze change by advocating for sustainable practices and holding industries accountable. In essence, the journey towards decarbonizing the global industrial landscape is a multifaceted endeavour that spans economic, environmental, and social dimensions, offering the potential not only to combat climate change but also to shape a more equitable, sustainable, and prosperous future [3, 4].

10.1.1 Decarbonization Initiative Options in Production Industries

Decarbonizing production industries is a critical endeavour aimed at curbing greenhouse gas emissions and mitigating the adverse effects of climate change. Production industries encompass diverse sectors, including manufacturing, mining, agriculture, and construction. Below are various crucial strategies and options for achieving decarbonization in these industries:

- *Transition to Renewable Energy*: A pivotal move involves shifting from conventional fossil fuel-based energy sources to renewable alternatives such as wind, solar, and hydroelectric power. This transition empowers industrial processes with clean electricity, leading to a substantial reduction in carbon emissions.
- *Enhancing Energy Efficiency*: An impactful approach entails improving energy efficiency within production facilities. This can be achieved

through the adoption of cutting-edge technologies, process optimization, and equipment upgrades, which minimize energy wastage.

- *Electrification processes*: In sectors where electrification is viable, replacing traditional combustion-based methods with electric alternatives can significantly lower emissions. Electric furnaces, heating systems, and machinery often exhibit enhanced efficiency and environmental friendliness.
- *Carbon Capture and Storage (CCS)*: For industries grappling with emissions that are challenging to eliminate entirely, CCS technologies prove invaluable. These systems capture carbon dioxide emissions and securely store them underground, thus preventing CO_2 from entering the atmosphere and contributing to reduced carbon footprints.
- *Fuel Switching*: Industries heavily reliant on high-carbon fuels like coal or oil can transition to lower-carbon alternatives, such as natural gas or hydrogen, to achieve emissions reductions.
- *Material Efficiency and Recycling*: Efforts to reduce material waste and promote recycling can effectively reduce emissions associated with resource extraction and production processes. Adopting a circular economy approach encourages the reuse and recycling of materials.
- *Optimizing Supply Chains*: Companies can scrutinize their supply chains to pinpoint opportunities for emissions reduction. Strategies such as local material sourcing, streamlined transportation routes, and waste minimization contribute significantly to decarbonization.
- *Carbon Pricing*: The implementation of a carbon pricing mechanism, such as a carbon tax or cap-and-trade system, serves as a powerful incentive for companies to reduce emissions by assigning a monetary cost to carbon emissions.
- *Green Procurement*: Encouraging the use of low-carbon or carbon-neutral products and materials in the production process can further slash emissions. This involves sourcing sustainable materials and components.
- *Sustainable Practices and Certifications*: Embracing sustainable practices and obtaining certifications like ISO 14001 for environmental management helps industries showcase their dedication to reducing their carbon footprint.

Decarbonizing maritime transport is intricately linked to the United Nations Sustainable Development Goals (SDGs), particularly Goal 13, "Climate Action." This connection is pivotal for advancing global environmental conditions through various means. Efforts in this context include building resilience and adaptive capacity in the face of climate-related hazards, integrating climate measures into national policies and maritime strategies, enhancing awareness and education on climate change mitigation and adaptation, and fostering international commitment and support for emissions reduction, especially in developing nations [5, 6]. These initiatives align with the SDGs' overarching objective of combatting climate change and promoting sustainable practices.

In the Asia-Pacific region, several prominent international organizations, including ASEAN, the Asian Development Bank (ADB), Asia Pacific Economic Cooperation (APEC), BIMSTEC (Bay of Bengal Initiative for Multi-Sectoral Technical and Economic Cooperation), and UNESCAP (United Nations Economic and Social Commission for Asia and the Pacific), have launched initiatives aimed at promoting awareness of green and low-carbon agendas. These initiatives serve as platforms for engaging governments and businesses as key stakeholders in advancing sustainable practices and decarbonization efforts. In the realm of maritime transportation, numerous nations adhere to the Greenhouse Gas Emission Goals set forth by the International Maritime Organization (IMO), aligning their strategies with international standards and obligations aimed at curtailing emissions from the shipping sector [7, 8]. This highlights the widespread agreement and cooperative efforts to reduce carbon footprints within a critical industry, thereby making substantial contributions to the broader objectives of climate mitigation and sustainability.

10.2 Eco-Friendly Energy Optimization: Strategies for Decarbonization

10.2.1 Demand-side Measures

Constitute a pivotal strategy in the quest for industrial sector decarbonization. The core idea revolves around actively reducing the demand for industrial products, thereby leading to decreased production and the consequent reduction in carbon emissions. This strategy not only aligns with sustainability objectives but also fosters resource efficiency. To exemplify this approach, the concept of "lightweighting" products, a deliberate practice focused on diminishing the necessity for materials such as steel and cement should be considered. By designing products to be lighter without sacrificing their functionality, industries can achieve substantial reductions in the consumption of resource-intensive materials. For instance, in the automotive sector, lighter vehicle designs translate to reduced demand for steel, leading to lower production-related emissions.

Furthermore, increasing the circularity of products through robust reuse and recycling programs is another effective means of diminishing industrial demand. Embracing a circular economy model involves extending the lifespan of products, components, and materials, thereby lessening the need for continuous manufacturing and resource extraction

[9, 10]. For instance, when electronics are designed for easier disassembly and recycling, it not only minimizes electronic waste but also curtails the production emissions associated with creating new devices. These initiatives not only contribute to reducing carbon emissions but also foster a more sustainable and resource-efficient industrial landscape, thus aligning with global efforts to combat climate change and promote environmental stewardship.

10.2.2 Energy Efficiency Improvements

One of the most impactful strategies in the pursuit of decarbonization across various industrial sectors is the implementation of energy efficiency improvements, which can substantially reduce fuel consumption for energy use by approximately 15 to 20%. This approach represents a win-win scenario, where industries can simultaneously cut their carbon footprint and operational costs. To exemplify this concept, let us delve into a few illustrative examples.

Firstly, energy-saving initiatives play a pivotal role in driving improvements in energy efficiency. These programs involve a comprehensive assessment of energy consumption patterns within industrial facilities, followed by the implementation of customized measures aimed at reducing waste and boosting energy productivity. For example, upgrading lighting systems to be more energy-efficient, modernizing HVAC systems, and optimizing production processes to minimize energy-intensive steps are all integral components of energy-saving programs [11, 12]. These endeavours not only lead to decreased fuel consumption but also result in significant cost savings over time.

Secondly, locating and removing non-value-added activities from industrial processes requires the deployment of Value Stream Mapping (VSM) approaches. With the use of VSM, the entire production or operational process may be thoroughly analyzed to spot inefficiencies like long wait times, surplus inventory, overproduction, waste production, double handling, and tasks that add no value to the finished product. Industries can considerably cut their energy use while also improving overall productivity by simplifying and optimizing these processes. For instance, production facilities can operate more effectively by minimizing unneeded downtime and inventory levels, which will result in decreases in both energy usage and related emissions.

Adopting energy efficiency upgrades is a key tactic for reducing fuel usage across different industrial sectors. Industries can significantly reduce their carbon emissions while also improving their competitiveness and sustainability in a global marketplace that is becoming more environmentally conscious by embracing energy-saving programs and using value stream mapping techniques to streamline operations.

10.2.3 Electrification of Heat

The electrification of heat represents a pivotal strategy in the quest for decarbonization, aiming to transition from fossil fuel-based heat generation to a zero-carbon electrification approach. This paradigm shift entails substantial changes in production processes and upfront investments, with the ultimate goal of reducing carbon emissions and fostering sustainability. To illustrate the concept, the electrification of ethylene production should be considered, a prime example of how this transformation can take place.

Traditional ethylene manufacturing typically relies on burning fossil fuels to generate high temperatures, which emits a considerable amount of carbon dioxide. This process will undergo a significant shift as a result of the addition of electric furnaces and electrically powered compressors. Electric furnaces are capable of achieving the necessary high temperatures using electricity instead of combustion, fundamentally changing the heat source. Concurrently, electrically driven compressors replace their fossil fuel-powered counterparts, further diminishing carbon emissions throughout the production chain. Nevertheless, this transition presents its set of challenges. Shifting to electric heat generation often requires substantial initial investments in new infrastructure and equipment, which can pose a barrier for some industries. Additionally, alterations in the production process may necessitate adjustments in operational procedures and workforce training.

Nonetheless, the advantages of electrification are remarkable. When applied to processes like ethylene production, the electrification of heat not only dramatically reduces carbon emissions, but also aligns with sustainability objectives and contributes to the global fight against climate change [13, 14]. It signifies a proactive stride toward a greener, more environmentally responsible future, where industries can flourish while significantly mitigating their carbon footprint. In essence, the electrification of heat emerges as a compelling strategy that underscores the pivotal role of innovation and transformative technologies in the transition toward a more sustainable industrial landscape.

10.2.4 Hydrogen Usage

Hydrogen can serve as an efficient means of storing excess renewable energy. During periods of surplus, electricity can be used to produce hydrogen through electrolysis. Leveraging hydrogen usage as a means to transition away from fossil fuel-based heat generation to a zero-carbon hydrogen approach is a transformative strategy that holds immense promise in the realm of decarbonization. This method entails substantial changes in production processes and requires substantial investments, all aimed at curbing carbon emissions and advancing sustainability. Two illustrative examples shed light on the potential of hydrogen in this context.

Firstly, ammonia production is a sector that traditionally relies on natural gas as a feedstock. To decarbonize this process, the game-changer is replacing the natural gas feedstock with zero-carbon hydrogen. This transformation involves a fundamental shift in the raw materials used, requiring substantial adjustments to the production infrastructure. By utilizing hydrogen, ammonia production can dramatically reduce its carbon footprint, aligning with sustainability objectives and significantly contributing to the global effort to combat climate change.

Second, another strong example of the application of hydrogen fuel cell technology is the development of hydrogen fuel cell automobiles. In order to power the electric motor in these vehicles, hydrogen is used as a clean energy source. This is accomplished by turning hydrogen gas into electricity. With this invention, there will be a paradigm shift away from fossil fuel-powered cars, effectively eliminating exhaust emissions and significantly lowering the transportation's carbon impact.

Nevertheless, transitioning to hydrogen-based solutions necessitates substantial investments, infrastructure modifications, and changes in the production process. These challenges can act as barriers for some industries and sectors. However, the potential rewards are significant. Hydrogen's ability to serve as a clean and versatile energy carrier makes it a cornerstone in the transition to a zero-carbon future [15, 16]. It not only offers a solution for decarbonizing industrial processes like ammonia production, but also presents a sustainable alternative for transportation, ultimately contributing to a greener, more environmentally responsible world. In essence, the adoption of hydrogen usage represents a visionary and pragmatic approach to decarbonization, where innovation and investments catalyze the transformation toward a more sustainable and climate-conscious global economy.

10.2.5 Biomass Usage

The utilization of biomass as an alternative to traditional fossil fuels signifies a momentous stride in the ongoing endeavour to decarbonize various industrial sectors. Biomass, encompassing solid forms such as wood and charcoal, liquid variants like biodiesel and bioethanol, and gaseous biogas, presents a versatile array of sustainable energy sources with transformative potential. Illustrative examples from diverse industries underscore the profound impact of biomass as a clean and renewable substitute for conventional fossil fuels.

In the steel production sector, a noteworthy shift has transpired as producers increasingly turn to charcoal derived from biomass. Charcoal serves as a dual-purpose solution by replacing coal as both a fuel source and a feedstock. This transition to charcoal marks a pivotal departure from the carbon-intensive coal usage traditionally associated with steel manufacturing processes. Notably, this innovative approach not only aligns with sustainability goals but also fosters the evolution of a more environmentally conscious and responsible steel production industry.

Similarly, the chemical sector is exploring the potential of naphtha, sourced from biomass, as a feedstock in chemical manufacturing processes. Bio naphtha offers a renewable and low-carbon alternative to conventional petrochemical feedstocks. By reducing its reliance on fossil fuels, the chemical industry effectively reduces its greenhouse gas emissions footprint while advancing sustainable practices in chemical production.

Moreover, Thailand's industrial landscape serves as a compelling case study, where biomass takes centre stage in powering diverse sectors [17, 18]. Industries in Thailand have harnessed the energy potential of biomass to fuel not only electricity generation but also paper and cement production. This multifaceted application of biomass showcases its versatility as a sustainable energy source with the capacity to significantly reduce the carbon footprint across various industrial domains.

However, the adoption of biomass necessitates prudent considerations, including sustainable sourcing practices, logistical challenges, and the development of efficient conversion technologies. Despite these challenges, the transition to biomass usage underscores a broader shift towards greener and more sustainable industrial practices. It exemplifies how the exploration and harnessing of renewable energy sources can play a pivotal role in mitigating climate change and advancing the global agenda for a sustainable, carbon-neutral future. In essence, biomass usage stands as a beacon of hope in the transition away from fossil fuels, offering not only a substantial reduction in carbon emissions but also a tangible pathway to more environmentally responsible and economically viable industrial processes.

10.2.6 Carbon Capture

CCUS, which stands for Carbon Capture, Utilization, and Storage, entails the capture of carbon dioxide (CO_2) emissions, typically from major point sources like industrial facilities and power plants utilizing fossil fuels or biomass as their primary energy source. Once captured, if not used on-site, CO_2 is compressed and transported via pipelines, ships, railways, or trucks for various applications. Alternatively, it can be injected deep into geological formations like depleted oil and gas reservoirs or saline aquifers for long-term storage.

One of CCUS's noteworthy advantages is its adaptability to existing power and industrial plants, facilitating their continued operation while significantly reducing emissions. This technology is particularly useful for reducing emissions in demanding industries, such as heavy ones like cement, steel, and chemicals. Additionally, CCUS is essential to the cost-effective production of low-carbon hydrogen, which in turn helps to decarbonize a number of different industries, transportation systems, and maritime commerce. Furthermore, CCUS offers the unique capability to extract CO_2 directly from the atmosphere, helping offset emissions that are either unavoidable or technically challenging to eliminate [19–21]. While the deployment of CCUS has fallen short of earlier expectations, recent years have witnessed a substantial surge in momentum, with more than 500 projects spanning the entire CCUS value chain currently in various stages of development. Nonetheless, even with this growth, CCUS deployment remains below what is deemed necessary to align with the ambitious goals of the Net Zero Scenario for mitigating climate change.

10.3 Policies

10.3.1 Carbon Pricing and Financial Incentives

10.3.1.1 Understanding Carbon Pricing and Its Mechanisms

Carbon pricing policies require emitters to pay a fee per ton of CO_2 or greenhouse gases (GHGs) they emit. This approach, which is technology-neutral, encourages cost-effective emissions reduction while also generating government revenue. Carbon pricing can be implemented through a carbon tax (fee per unit of emissions) or a cap-and-trade system (emissions permits purchased in a market). In some cases, a hybrid approach combining cap-and-trade with price ceilings and floors can effectively balance economic and environmental objectives [22].

10.3.1.2 Realizing Benefits and Addressing Leakage Concerns

Carbon pricing brings several advantages, such as facilitating efficient responses across economic channels, including input substitution, process changes, and demand reduction. Notably, regulators do not require in-depth industry knowledge to set emissions standards, making it a practical tool for promoting cost-effective emissions reduction. Additionally, government revenue generated through carbon pricing can support socially beneficial programs or tax reductions. To address leakage concerns, which involve industrial activities relocating to regions with weaker or absent carbon pricing, various approaches can be employed [23, 24]. These include the free allocation of emissions allowances to energy-intensive, trade-exposed industries, tax rebates based on output, and the implementation of border tax adjustments.

10.3.1.3 Holistic Policy Approaches and Enabling Technologies

To foster emissions reductions in the industry sector, it is imperative to adopt a combination of policies that collectively promote decarbonization across various economic channels. Tradable intensity performance standards, such as vehicle fleet efficiency standards and low-carbon fuel standards, provide incentives for input substitution and process changes without limiting output. However, no single policy can serve as a "silver bullet" for industry decarbonization. Instead, a comprehensive package of policies is necessary, with components like carbon pricing, emissions standards, R&D support, government procurement, labelling, and disclosure requirements all playing pivotal roles. These policies work in synergy to drive emissions reductions. Moreover, policies like R&D support may not independently achieve emissions reductions across all channels but serve to enable the development of cost-effective emission-reducing technologies. They act as bridges between laboratory research and market implementation. For governments, creating demand through procurement policies is a vital aspect of supporting low-carbon technology development and commercialization [25]. This holistic approach and the collaboration of policies across various fronts are critical policy imperatives for achieving industry sector decarbonization and a sustainable future.

10.3.1.4 Government Procurement Driving Sustainability and Emissions Reduction

Government procurement practices worldwide have demonstrated their substantial impact on advancing sustainability and reducing emissions. Successful programs such as the Buy Clean California Act, Japan's Act on Promoting Green Purchasing, and India's Ujala program for efficient lighting have resulted in significant emissions reductions and energy savings. These examples underscore the effectiveness of government procurement in achieving sustainability goals.

Italy and Sweden have taken innovative approaches to government procurement to promote sustainability and emissions reduction. In Italy, Green Public Procurement (GPP) employs Minimum Environmental Criteria (CAM) and eco-labels, aiming to enhance sustainability while requiring capacity building and support for procurement entities. Sweden, on the other hand, actively encourages the exploration of new Sustainable Public Procurement (SPP) methods, driven by existing tools and supplier engagement. Notably, Sweden's transport and building sectors have led in low-carbon procurement [25–27]. Despite their achievements, both countries face challenges such as risk aversion and competition in the competitive low-carbon marketplace. These initiatives underscore the pivotal role of government procurement in advancing sustainability and emissions reduction endeavours.

10.4 Research Development and Demonstration (RDD) Support

10.4.1 The Role of RD&D in Industrial Innovation and Collaboration

Research, Development, and Demonstration (RD&D) efforts are pivotal in driving industrial innovation, addressing scaling challenges, and advancing technologies. These challenges manifest across different stages of market development, necessitating coordination between various manufacturing scales. Furthermore, innovation flourishes as companies transition from prototyping and demonstration phases to full-scale commercialization. Collaboration is integral to this process, with research partnerships involving government, industry, and philanthropic organizations, collectively fostering technological advancements and addressing challenges. Initiatives such as Canada's Sustainable Development Technology Canada, the U.S. ARPAE, and successful collaborations like Elysis exemplify how industry, government, and philanthropy can work together to overcome technological obstacles and drive innovation in industrial sectors. Additionally, philanthropic investments from organizations like PRIME Coalition [28, 29]. Breakthrough Energy Ventures provide crucial support for long-term RD&D efforts in industrial technologies.

10.4.2 Fostering Industrial RD&D through Policy Support

Government policies play a critical role in nurturing industrial Research, Development, and Demonstration (RD&D) efforts across various stages of technology maturity. These policies encompass a spectrum of initiatives, including government laboratories such as the U.S. DOE Labs and France's CNRS Institutes, which provide invaluable expertize and research facilities. Additionally, government incentives like tax credits and grants serve as powerful catalysts, incentivizing corporations to invest in RD&D activities and accelerate technological advancements.

To ensure the effectiveness of RD&D programs and policies, several fundamental elements come into play:

1. *Collaborative Partnerships*: Partnering with companies at all stages of market development promotes knowledge sharing, fosters innovation, and expedites the adoption of new technologies.

2. *Targeted RD&D*: Concentrating RD&D efforts on addressing specific technical challenges unique to the industry ensures that resources are allocated efficiently, driving tangible results.
3. *Involvement of Trade Associations and Government*: Engaging trade associations and government entities in the RD&D process ensures that emerging technologies are readily accessible to multiple companies, maximizing their impact.
4. *Promoting Innovation and Adoption*: Encouraging innovation and the swift adoption of newly developed solutions are pivotal in propelling progress within industrial RD&D, driving sustainable advancements in various sectors [30–32].

10.5 Energy Efficiency or Emissions Standards

10.5.1 Maximizing Energy Efficiency for Sustainable Progress

Energy efficiency standards, applied across appliances, equipment, and vehicles, have proven highly effective in reducing energy consumption in both developed and developing nations. Electric motor standards, in particular, have played a pivotal role in curbing industrial electricity usage. However, there exists untapped potential for even greater energy efficiency within the industrial sector. Numerous cost-effective energy efficiency upgrades promise rapid financial returns, often with payback periods of less than two years. Regrettably, manufacturers occasionally miss these opportunities due to perceived economic disincentives and budget constraints. A more comprehensive evaluation of energy efficiency enhancements, considering factors beyond energy savings—such as increased productivity, reduced environmental impact, enhanced safety measures, and other advantages—holds the potential to reshape investment decisions [33–35]. This broader perspective extends to non-energy benefits, including revenues from emission reduction credits, competitive market advantages, lowered maintenance costs, and more.

10.5.2 Promoting Energy Efficiency and the Role of Standards

Promoting energy efficiency poses challenges when businesses prioritize other cost-saving measures over investing in upgrades, particularly when energy costs are a relatively small part of their expenses. Efficiency standards have emerged as key drivers of energy-saving initiatives across various industries. These standards, once predominantly technology-specific, have evolved toward performance-oriented benchmarks. They have demonstrated their effectiveness, particularly for mass-produced products, usually applied at the point of purchase or installation. Retrofitting, however, remains a challenge due to compliance issues [36, 37].

Now, the industry is transitioning towards system-level standards, specifying acceptable energy consumption for systems providing specific services. These standards are most effective for applications with well-defined systems, widespread usage, stable performance indicators, and collaboration among key stakeholders [35, 38]. Examples of potential targets include systems like steam and hot-water systems, dryers, water treatment systems, chilled water systems, and cooling towers. These performance standards complement carbon pricing efforts, maximizing emissions reduction.

10.6 Building Codes and Standards

10.6.1 Promoting Sustainable Building Practices

Buildings are essential for human well-being, yet they account for a substantial portion of global energy consumption and process-related greenhouse gas (GHG) emissions, particularly during their construction phase. The production of structural materials like steel and concrete alone contributes to more than 20% of the world's energy and process-related emissions. To mitigate these environmental impacts, strategies are being employed to reduce emissions in construction. These include optimizing material use, extending the performance life of buildings, and improving material composition.

However, the complexity of design guidelines for sustainable buildings can pose challenges. These guidelines may address different aspects of a building's lifecycle, from initial design choices to retrofitting, often focusing on individual components or specific moments in time. This approach can inadvertently lead to unintended consequences for subsequent components or phases of a building's lifecycle [36, 39]. Balancing trade-offs between different materials and making minor alterations in construction scheduling can have significant environmental implications.

10.6.2 Enhancing Building Sustainability Through Materials and Codes

A sustainable approach to building design and construction involves several key considerations. First, the selection of long-lived materials for interior finishes and roofing is crucial, especially as many other building materials have shorter lifespans. Durable materials contribute to the reduction of the environmental impact of buildings [37, 40]. Furthermore, the prefabrication of building systems is gaining prominence as a sustainable construction approach. Prefabrication not only reduces waste and on-site construction time, but also enhances overall quality and durability, ultimately extending the lifespan of buildings.

This approach is particularly valuable in minimizing waste, a critical aspect in the context of developing economies. In addition to materials and construction methods, the modernization of building codes is essential to keep up with emerging technologies and materials in the construction industry. Modern building codes can promote sustainability by, for example, encouraging the use of supplementary cementitious materials and reducing the cement content in concrete. These updated codes, when coupled with clear sustainability targets, labelling systems, and incentives for alternative materials, play a vital role in facilitating the integration of environmentally friendly practices and materials into building construction and design [41, 42].

10.7 Electrification and Fuel Switching

10.7.1 Hydrogen as a Versatile Energy Carrier and Decarbonization Tool

Hydrogen, a versatile energy carrier, holds significant promise in the transition to a low-carbon future. It is one of several potential energy carriers derived from hydrogen, alongside ammonia, methane, and methanol. Among these, ammonia and methane stand out due to their advantages in terms of transport and the utilization of existing infrastructure. Notably, both ammonia and methane can be efficiently produced from hydrogen, making low-cost, carbon-free hydrogen production a valuable asset in the quest for sustainability. Currently, the global hydrogen landscape is dominated by production methods, with 76% of hydrogen sourced from methane steam reforming, 22% from coal gasification, and a mere 2% from electrolysis. However, there is a growing focus on decarbonizing hydrogen production, particularly through methods like steam methane reforming coupled with carbon capture, resulting in what is often referred to as "blue" hydrogen [43, 44]. The cost-effectiveness of "blue" hydrogen production varies by the region, and emerging technologies, such as methane pyrolysis with carbon co-products, show promise in further reducing costs.

Due to its adaptability across a range of uses, hydrogen is an important tool in the transition to a sustainable energy future. It can be utilized as a chemical feedstock, in high-temperature industrial operations, building heating systems, fuel cell electric vehicles, and heating systems. Notably, hydrogen offers a sustainable substitute in the manufacturing of ammonia for fertilizers and various chemicals [45]. Additionally, hydrogen can help with the integration of variable electricity generation through energy storage, improving grid stability and the use of renewable energy.

10.7.2 Challenges and Considerations in Hydrogen Energy

Hydrogen has a special set of difficulties and factors to take into account when using it as an energy carrier, which affects how well-suited it is for different uses. Hydrogen has a low energy density relative to thermal fuels, which is the main reason for its first-world energy storage problems. This restriction may limit its applicability, especially in situations involving long-distance flying, heavy freight transport, industrial processes involving high temperatures, and specific chemical feedstocks, where other energy carriers may be more appropriate choices. There are unique difficulties with coal-based hydrogen generation, which are more common in places like China, when it comes to the creation of hydrogen [46]. This method is carbon-intensive and encounters difficulties in carbon capture due to impurities, highlighting the need for cleaner and more sustainable hydrogen production pathways.

Alkaline electrolysis, Solid Oxide Fuel Cells (SOFCs), and Proton Exchange Membrane FuelCells (PEMFCs) are a few of the several electrolysis methods that are available. Electrolysis is a promising way for producing hydrogen. Although SOFCs and PEMFCs are more efficient and economical, they have issues with capital expenditures, electricity costs, and water utilization. Hydrogen's storage presents another hurdle, given its inherently low density.

Existing storage methods include compression and liquefaction, both of which are energy-intensive processes. Additionally, solid-state storage options exist but are often constrained by limited accessibility. Transporting hydrogen via trucks or pipelines is a feasible approach, but it requires the development of new infrastructure. Furthermore, hydrogen's small molecular size makes it susceptible to leaks, necessitating specialized transport solutions [47]. Ensuring safety and addressing regulatory harmonization are essential components of the evolving hydrogen transportation landscape. Navigating these challenges and considerations is crucial to harnessing hydrogen's potential as a versatile and sustainable energy carrier in our transition to a cleaner and more efficient energy future.

10.7.3 Electrification for a Sustainable Future

Electrification emerges as a central pillar in the global pursuit of sustainability and the reduction of greenhouse gas emissions. This transformative approach involves the widespread adoption of electric power across sectors traditionally reliant on fossil fuels, offering the potential to revolutionize transportation, heating, and industrial processes, thereby paving the way for a cleaner and more environmentally friendly energy future. In the current landscape of industrial energy use, direct fuel combustion remains dominant, with electricity playing a secondary role. Fuels are predominantly deployed for tasks such as process heating and boiler operations. Efforts to realize electrification encompass a diverse array of technologies, including heat pumps, induction heating, and electromagnetic waves. These innovations open up specific applications amenable to electrification, such as laser sintering, resistive heating, and electric arc furnaces.

However, electrification faces a significant obstacle in the form of cost competitiveness, necessitating a comprehensive evaluation of efficiency and heating requirements, particularly concerning thermal fuels [47–49]. Addressing the challenges posed by industrial electrification entails grappling with issues of application specificity, equipment design, scale-up or down considerations, and the interconnected nature of relevant technologies. An integrated approach to research and development is indispensable in surmounting these challenges, ensuring a successful transition towards a sustainable and electrified future.

10.8 Circular Economy Practices

10.8.1 Embracing the Circular Economy for Sustainable Practices

The Circular Economy (CE) paradigm represents a transformative concept centred on optimizing the value of products, components, or materials through a spectrum of eco-conscious practices. These include but are not limited to reuse, resource sharing, refurbishment, remanufacturing, and recycling, fundamentally diverging from the traditional linear economy where products meet their end as waste. Central to the realization of a Circular Economy are key strategies that encompass the extension of product lifespans, the promotion of product sharing, the embracing of refurbishing and remanufacturing processes, and the cultivation of efficient recycling systems. A cornerstone of these strategies lies in the meticulous maintenance and enhancement of product information, ensuring traceability and informed decisions throughout a product's lifecycle.

Circular Economy brings forth a multitude of benefits, including substantial reductions in energy consumption, greenhouse gas (GHG) emissions, and overall materials demand. However, it is not without challenges [50–52]. Achieving comparable performance with virgin materials stands as a pivotal obstacle, necessitating innovation and collaboration. Furthermore, the effective tracking of information pertaining to materials and components plays a critical role in the Circular Economy's success. To foster a Circular Economy successfully, proactive design for reuse and recycling emerges as an essential approach. This encompasses the integration of modular design principles, the adoption of reversible attachments, and the standardization of materials—pillars that lay the foundation for sustainable circularity.

10.8.2 Enabling Circular Economy through Thoughtful Policies

Unlocking the potential of Circular Economy (CE) requires a well-structured policy framework that guides and encourages sustainable practices. One pivotal policy intervention is the implementation of Extended Producer Responsibility (EPR), which holds manufacturers accountable for managing the end-of-life phase of their products. This move incentivizes greater consideration of product durability and recyclability in the design phase. Complementary measures include the adoption of leasing models that promote sharing and extended use of products, the establishment of reverse supply chains for efficient product recovery, the imposition of product disassembly requirements to facilitate recycling, and the development of robust information infrastructure to monitor and optimize material flows throughout the CE ecosystem.

Embracing the Circular Economy is not merely a choice but an imperative. It is a cornerstone in the collective efforts to combat global warming, reduce the carbon intensity of material consumption, and uphold living standards and sustainable development [52–54]. To enact the principles of the Circular Economy effectively, a multifaceted approach in various policy areas is essential. This encompasses the enhancement of building codes to promote sustainable construction, the establishment of zoning policies that facilitate shared mobility, the creation of reverse supply chains, the enforcement of product disassembly requirements, and the advancement of information infrastructure for real-time tracking of material flows. These combined endeavours pave the way for a sustainable and circular future.

10.9 Challenges in Scaling CCUS for Net-Zero Goals

10.9.1 The Urgent Need to Accelerate Carbon Capture, Utilization, and Storage (CCUS)

While there have been notable strides in Carbon Capture, Utilization, and Storage (CCUS) developments, the progress remains insufficient to align with global net-zero emission goals. Even if all planned CCUS projects were successful, they would capture less than 15% of the volume required according to the IEA Net Zero Scenario. This situation highlights several critical challenges, including fierce competition for funding, a reliance on policy support, and the imperative to expedite the development and permitting processes for CO_2 storage resources. CCUS plays a pivotal role in the broader challenge of transitioning to a net-zero emissions future. It is one of the 46 crucial technologies assessed in the IEA's 'Tracking

Clean Energy Progress' analysis, with only electric vehicles and lighting considered to be on track [55, 56]. This underscores the unprecedented scale and pace required for the global energy transition to achieve net-zero emissions by 2050. Recent national net-zero commitments have rekindled interest in CCS, recognizing its essential role in achieving substantial carbon abatement.

A fundamental starting point for CCS often involves capturing CO_2 from industrial effluent gas streams, with an initial focus on high-concentration streams, such as those originating from biofuels or ammonia plants. These high-concentration streams offer lower capture costs, making the CCS business model more economically feasible. The presence of anchor tenants with high-concentration streams can help establish essential infrastructure for the transportation and storage of captured CO_2, particularly in hubs or clusters. The quest for effective CCS extends to tackling dilute streams, presenting a considerable challenge. Achieving commercial viability for CCS in dilute streams requires the reduction of capture costs to approximately US$30 per ton or lower. Emerging technologies like electro-swing adsorption, metal-organic frameworks, and solid sorbents show promise but necessitate further development [56–58]. Enhancing the efficiency of traditional amine solvents and creating regional supply chains can also contribute to lowering operating costs. These efforts are imperative in expediting progress towards comprehensive CCUS implementation.

10.9.2 Innovations in Carbon Removal and Storage

As the world grapples with the imperative of reducing carbon emissions, innovations in carbon removal and storage are becoming increasingly significant. Among these innovations is "Direct Air Capture," a technology that allows for the capture of CO_2 directly from the atmosphere. This approach is particularly valuable for addressing emissions from diffuse and hard-to-reach sources. However, to facilitate large-scale adoption, it is crucial to drive down the costs associated with Direct Air Capture and establish effective carbon trading mechanisms.

Another critical aspect of carbon management involves addressing the challenges related to the sequestration and long-term storage of CO_2. While CO_2 has been utilized for enhanced oil recovery in the past, achieving net-zero emission goals necessitates the permanent sequestration of larger volumes of CO_2 in deep saline formations [59–61]. This endeavour requires rigorous site selection and permitting processes to ensure the safe and enduring storage of carbon dioxide. Embracing hub-based approaches can further enhance the feasibility of carbon storage by spreading the responsibility for storing larger volumes of CO_2 across multiple sites, thereby reducing unit costs and associated risks. These innovations and strategies are pivotal in advancing our efforts to combat climate change and achieve net-zero emissions.

10.9.3 Navigating Regulatory Challenges and Fostering CCS as a Climate Solution

The deployment of Carbon Capture and Storage (CCS) faces formidable regulatory and permitting challenges, often leading to delays in storage site development. The lack of regulatory maturity and understanding contributes to this issue, exemplified by the lengthy process of obtaining a Class VI injection permit for dedicated storage sites in the US. These hurdles, coupled with the high development costs and liability associated with carbon storage, can deter large-scale deployment.

However, CCS presents a unique climate change mitigation solution. Unlike renewables that provide tangible products or power, CCS mitigates climate change while allowing continued production with a significantly reduced carbon footprint. Realizing the full potential of CCS hinges on the creation of mechanisms that generate value for decarbonization efforts, be it through policies, regulations, or market incentives [60–63]. Collaborative partnerships among organizations spanning various aspects of CCS projects are essential for success.

The path forward for CCS involves a crucial transition from discovery and feasibility studies to development and execution. This shift requires the support of attractive policies and consumer valuation of low-carbon products. Flagship CCS projects in strategic geographies will lay the foundation for extensive CCS networks, transforming it into a thriving business with immediate demand and substantial investments. To scale CCUS effectively, policies should establish sustainable and viable markets for investment, tailored to the specific needs of different CCUS applications. Industrial CCUS clusters that share infrastructure can achieve economies of scale, enabling the capture of CO_2 at numerous smaller industrial facilities. Government support for early planning and coordination of CO_2 transport and storage infrastructure can reduce lead times for future CCUS facilities [64–66]. Additionally, proactive development of CO_2 storage resources is imperative to meet the rising demand for CCUS.

10.10 International Climate Agreements

10.10.1 The Role of International Climate Agreements

- *Paris Agreement*: The Paris Agreement, established in 2015, stands as a pivotal international climate accord. Its objective is to cap global warming at well below 2 degrees Celsius above pre-industrial levels, with a more ambitious target of limiting it to 1.5 degrees. Industrial decarbonization holds a crucial position in realizing these goals.

- *Nationally Determined Contributions (NDCs)*: Within the Paris Agreement framework, participating nations present their Nationally Determined Contributions (NDCs), outlining their commitments to reducing emissions. A noteworthy trend is the inclusion of strategies to cut industrial emissions in many countries' NDCs.
- *Technology Transfer and Financing*: The Paris Agreement encourages global collaboration in technology transfer and provides financial mechanisms to support developing nations in their endeavours to decarbonize their industrial sectors.
- *Montreal Protocol*: Initially focused on phasing out ozone-depleting substances, the Montreal Protocol serves as a valuable model for demonstrating how international agreements can effectively stimulate technological innovation. It provides lessons for addressing emissions of industrial gases with high global warming potential, like hydrofluorocarbons (HFCs).
- *Phasedown of HFCs*: The Kigali Amendment to the Montreal Protocol outlines a gradual reduction in the production and use of HFCs, commonly employed in air conditioning and refrigeration systems. This initiative significantly contributes to diminishing emissions originating from industrial sources.
- *International Energy Agency (IEA)*: The IEA plays a pivotal role in facilitating international collaboration and the exchange of best practices for achieving industrial decarbonization. It conducts comprehensive analyses and provides policy recommendations to its member countries.
- *Energy Efficiency and Technology Roadmaps*: The IEA offers guidance on enhancing energy efficiency within the industrial sector and formulating technology roadmaps to steer the transition toward decarbonization [67–70].

10.10.2 Advancing Carbon Capture and Storage (CCS) Across European Nations

Several European nations are taking significant steps to advance Carbon Capture and Storage (CCS) as a critical component of their climate strategies. In the European Union (EU), the 'Fit for 55' Package introduced in July 2021 aims to align EU policies with a 55% reduction in greenhouse gas emissions by 2030. A key aspect of this initiative is the proposal to include all CO_2 transport modalities under the EU Emissions Trading System (ETS) Directive. Furthermore, the EU Commission's revision of the TEN-E regulation underscores the importance of aligning all CO_2 transport and storage modalities with the EU's climate neutrality goals. The EU recognizes the 2020s as a pivotal decade for achieving climate neutrality, with CCS playing a central role in these efforts.

In the United Kingdom (UK), the focus has primarily been on decarbonizing the electricity sector, resulting in significant emissions reductions. However, this approach has brought challenges related to the intermittency of renewable sources and the rise of microgeneration, impacting the stability of the electricity market. As the UK progresses toward its net-zero emissions target, the challenges will extend to decarbonizing the transport and heating sectors, potentially necessitating increased use of electricity, CCUS, and hydrogen technologies [71, 76]. Yet, uncertainties persist regarding the volume, usage, and production methods of hydrogen, as well as infrastructure development, which could affect the pace of progress.

Norway is demonstrating a strong commitment to CCS through its authorization of the Longship Project, backed by substantial government funding in the 2021 state budget. This project is expected to yield significant socio-economic benefits, potentially generating up to 5,000 jobs in Norway and positioning Norwegian companies as leaders in CCS due to their technical expertize.

Beyond economic gains, the project holds substantial technological and knowledge distribution potential, contributing to the development of future CO_2 management programs. Norway's CCS experience highlights the importance of prioritizing health, safety, and environmental aspects, fostering commercial innovation through models like Carbon Capture as a Service (CCaaS), and leveraging innovative funding mechanisms, such as the EU Innovation Fund (IF), to support CCS projects [77–81]. These initiatives underscore the collective commitment of European nations to harness CCS as a vital tool in their climate mitigation strategies.

10.11 Conclusion

This chapter discusses the various policies that have been implemented for decarbonization. The urgent need to decarbonize the global industrial sector has become increasingly apparent in the battle against climate change. Industries are major contributors to greenhouse gas emissions, making it absolutely essential to transform them for the sake of a sustainable future. Achieving decarbonization is a complex process that involves several key elements: transitioning to renewable energy sources, improving energy efficiency, adopting carbon capture and storage technologies, and electrifying industrial operations. Governments, industry, and society must work closely together on this project, which must also be supported by international cooperation, clean technology investments, and supportive regulations.

References

[1] Barker, Terry and Crawford-Brown, Douglas. Dec 2014. Decarbonising the world's economy: assessing the feasibility of policies to reduce greenhouse gas emissions.
[2] Stern, N. 2008. The economics of climate change. American Economic Review 98: 1–37.
[3] Tukker, A. and Dietzenbacher, E. 2013. Global Multiregional Input–Output Frameworks: An Introduction and Outlook. https://doi.org/10.1080/09535314.2012.761179 25: 1–19.
[4] AR5 Climate Change 2014: Mitigation of Climate Change — IPCC. https://www.ipcc.ch/report/ar5/wg3/.
[5] Åhman, M., Nilsson, L. J. and Johansson, B. 2016. Global climate policy and deep decarbonization of energy-intensive industries. http://dx.doi.org/10.1080/14693062.2016.1167009 17: 634–649.
[6] Abbasi, T. and Abbasi, S. A. 2011. Decarbonization of fossil fuels as a strategy to control global warming. Renewable and Sustainable Energy Reviews 15: 1828–1834.
[7] Francisco Lallana, Gonzalo Bravo, Gaëlle Le Treut, Julien Lefèvre, Gustavo Nadal and Nicolás Di Sbroiavacca. 2021. Exploring deep decarbonization pathways for Argentina. Energy Strategy Reviews 36: 100670.
[8] Dylan D. Furszyfer Del Rio, Benjamin K. Sovacool, Aoife M. Foley, Steve Griffiths, Morgan Bazilian, Jinsoo Kim and David Rooney. 2022. Decarbonizing the glass industry: A critical and systematic review of developments, sociotechnical systems and policy options. Renewable and Sustainable Energy Reviews 155: 111885.
[9] Mundaca, L., Ürge-Vorsatz, D. and Wilson, C. 2019. Demand-side approaches for limiting global warming to 1.5°C. Energy Effic 12: 343–362.
[10] Shogo Sakamoto, Yu Nagai, Masahiro Sugiyama, Shinichiro Fujimori, Etsushi Kato, Ryoichi Komiyama et al. 2021. Demand-side decarbonization and electrification: EMF 35 JMIP study. Sustain Sci. 16: 395–410.
[11] Hannah Förster, Katja Schumacher, Enrica De Cian, Michael Hübler, Ilkka Keppo, Silvana Mima et al. 2013. European Energy Efficiency and Decarbonization Strategies Beyond 2030—A Sectoral Multi-Model Decomposition. https://doi.org/10.1142/S2010007813400046 4.
[12] Mier, M. and Weissbart, C. 2020. Power markets in transition: Decarbonization, energy efficiency, and short-term demand response. Energy Econ. 86: 104644.
[13] Jing, R., Zhou, Y. and Wu, J. 2022. Electrification with flexibility towards local energy decarbonization. Advances in Applied Energy 5: 100088.
[14] Zuberi, M. J. S., Chambers, J. and Patel, M. K. 2021. Techno-economic comparison of technology options for deep decarbonization and electrification of residential heating. Energy Effic 14: 1–18.
[15] Maestre, V. M., Ortiz, A. and Ortiz, I. 2021. Challenges and prospects of renewable hydrogen-based strategies for full decarbonization of stationary power applications. Renewable and Sustainable Energy Reviews 152: 111628.
[16] Oshiro, K. and Fujimori, S. 2022. Role of hydrogen-based energy carriers as an alternative option to reduce residual emissions associated with mid-century decarbonization goals. Appl. Energy 313: 118803.
[17] Su, Y., Hiltunen, P., Syri, S. and Khatiwada, D. 2022. Decarbonization strategies of Helsinki metropolitan area district heat companies. Renewable and Sustainable Energy Reviews 160: 112274.

[18] Nunes, L. J. R. and Matias, J. C. O. 2020. Biomass torrefaction as a key driver for the sustainable development and decarbonization of energy production. Sustainability 12: 922.
[19] Herraiz, L., Lucquiaud, M., Chalmers, H. and Gibbins, J. 2020. Sequential combustion in steam methane reformers for hydrogen and power production with CCUS in decarbonized industrial clusters. Front Energy Res. 8: 539843.
[20] Frontiers | Sequential Combustion in Steam Methane Reformers for Hydrogen and Power Production With CCUS in Decarbonized Industrial Clusters. https://www.frontiersin.org/articles/10.3389/fenrg.2020.00180/full.
[21] Kawai, E., Ozawa, A. and Leibowicz, B. D. 2022. Role of carbon capture and utilization (CCU) for decarbonization of industrial sector: A case study of Japan. Appl. Energy 328: 120183.
[22] Carbon valuation in UK policy appraisal: a revised approach - GOV.UK. https://www.gov.uk/government/publications/carbon-valuation-in-uk-policy-appraisal-a-revised-approach.
[23] Blanchflower, D. G. and Oswald, A. J. 1995. An introduction to the wage curve. Journal of Economic Perspectives 9: 153–167.
[24] Pearce, D., Pearce and David. 1991. The role of carbon taxes in adjusting to global warming. Economic Journal 101: 938–48.
[25] González-Eguino, M. 2011. The importance of the design of market-based instruments for CO_2 mitigation: An AGE analysis for Spain. Ecological Economics 70: 2292–2302.
[26] Clinch, J. P. and O'Neill, E. 2010. Designing Development Planning Charges: Settlement Patterns, Cost Recovery and Public Facilities. http://dx.doi.org/10.1177/0042098009357968 47: 2149–2171.
[27] Šauer, P., Vojáček, O., Klusák, J. and Zimmermannová, J. 2011. Introducing Environmental Tax Reform: The Case of the Czech Republic. Environmental Tax Reform (ETR): A Policy for Green Growth. doi:10.1093/ACPROF:OSO/9780199584505.003.0006.
[28] Suter, F., Steubing, B. and Hellweg, S. 2017. Life cycle impacts and benefits of wood along the value chain: the case of Switzerland. J. Ind. Ecol. 21: 874–886.
[29] Energy Analysis, Data, and Reports | Department of Energy. https://www.energy.gov/eere/iedo/energy-analysis-data-and-reports.
[30] Press Releases | Alcoa Corporation. https://news.alcoa.com/press-releases/default.aspx.
[31] U.S. Research and Development Tax Credit: A First Look at the Effect of the PATH Act - ProQuest. https://www.proquest.com/docview/2225774793?pq-origsite=gscholar&fromopenview=true.
[32] Russell, C. 2015. Multiple benefits of business-sector energy efficiency: a survey of existing and potential measures.
[33] Davis, S. J. and Caldeira, K. 2010. Consumption-based accounting of CO_2 emissions. Proc. Natl. Acad. Sci. U S A 107: 5687–5692.
[34] John Barrett, Glen Peters, Thomas Wiedmann, Kate Scott, Manfred Lenzen, Katy Roelich et al. 2013. Consumption-based GHG emission accounting: a UK case study. Climate Policy 13: 451–470.
[35] Rose, A. and Chen, C. Y. 1991. Sources of change in energy use in the U.S. economy, 1972–1982: A structural decomposition analysis. Resources and Energy 13: 1–21.
[36] Bentzen, J. 2004. Estimating the rebound effect in US manufacturing energy consumption. Energy Econ. 26: 123–134.
[37] Allwood, J. M., Ashby, M. F., Gutowski, T. G. and Worrell, E. 2011. Material efficiency: A white paper. Resour. Conserv. Recycl. 55: 362–381.
[38] Hernández, R., Martínez Piva, J. M., Mulder, N. and BMZ. 2014. Global value chains and world trade: Prospects and challenges for Latin America [Cadenas globales

de valor y comercio mundial: Perspectivas y desafíos para América Latina]. OECD Science, Technology and Industry Working Papers 71.
[39] Beucker, S., Bergesen, J. D. and Gibon, T. 2016. Building energy management systems: global potentials and environmental implications of deployment. J. Ind. Ecol. 20: 223–233.
[40] Structure, C. M. and 2016, undefined. 2016. 3D printed structures: challenges and opportunities. structuremag.org.
[41] Bie A. Miller, Arpad Horvath, Monteir, P. J. M. and Claudia Ostertag. 2015. Greenhouse gas emissions from concrete can be reduced by using mix proportions, geometric aspects, and age as design factors. Environmental Research Letters 10: 114017.
[42] Kurtis, K. E. 2015. Innovations in cement-based materials: Addressing sustainability in structural and infrastructure applications. MRS Bull 40: 1102–1108.
[43] He, G., Mallapragada, D. S., Bose, A., Heuberger-Austin, C. F. and Gençer, E. 2021. Sector coupling via hydrogen to lower the cost of energy system decarbonization. Energy Environ. Sci. 14: 4635–4646.
[44] Yimin Deng, Raf Dewil, Lise Appels, Flynn Van Tulden et al. 2022. Hydrogen-enriched natural gas in a decarbonization perspective. Fuel 318: 123680.
[45] Kazi, M. K., Eljack, F., El-Halwagi, M. M. and Haouari, M. 2021. Green hydrogen for industrial sector decarbonization: Costs and impacts on hydrogen economy in qatar. Comput. Chem. Eng. 145: 107144.
[46] Parra, D., Gillott, M. and Walker, G. S. 2014. The role of hydrogen in achieving the decarbonization targets for the UK domestic sector. Int. J. Hydrogen Energy 39: 4158–4169.
[47] Wei, M., McMillan, C., Reports, S. de la R. du C.-/Renewable E. and 2019, undefined. 2019. Electrification of Industry: Potential, Challenges and Outlook. Springer 6: 140–148.
[48] Prussi, M., Laveneziana, L., Testa, L., Energies, D. C. and 2022, undefined. 2022. Comparing e-Fuels and Electrification for Decarbonization of Heavy-Duty Transports. mdpi.com. doi:10.3390/en15218075.
[49] Morte, I., Queiroz, F. O. de, ... C. M.-E. S. and 2023, undefined. 2023. Electrification and Decarbonization: A Critical Review of Interconnected Sectors, Policies, and Sustainable Development Goals. Elsevier.
[50] Zhang, C., Zhao, X., Sacchi, R. and You, F. 2023. Trade-off between critical metal requirement and transportation decarbonization in automotive electrification. Nature Communications 14(1): 1–16.
[51] Dustin Mulvaney, Ryan M. Richards, Morgan D. Bazilian, Erin Hensley, Greg Clough and Seetharaman Sridhar. 2021. Progress towards A Circular Economy in Materials to Decarbonize Electricity and Mobility. Elsevier.
[52] Energy, P. F.-A. and 2022, undefined. Analysing the systemic implications of energy efficiency and circular economy strategies in the decarbonisation context. aimspress. com 10: 191–218.
[53] Bleischwitz, R., Yang, M., Huang, B., Recycling, X. X.-... and 2022, undefined. The Circular Economy in China: Achievements, Challenges and Potential Implications for Decarbonisation. Elsevier.
[54] Koide, R., Murakami, S., Energy, K. N.-R. and S. & 2022, undefined. Prioritising Low-Risk and High-Potential Circular Economy Strategies for Decarbonisation: A Meta-Analysis on Consumer-Oriented Product-Service Systems. Elsevier.
[55] Nußholz, J., Çetin, S., Eberhardt, L., Advances, C. D. W.-... & R. and 2023, undefined. From Circular Strategies to Actions: 65 European Circular Building Cases and Their Decarbonisation Potential. Elsevier.
[56] Joule, J. D. and 2020, undefined. 2020. Is net zero carbon 2050 possible? cell.com. doi:10.1016/j.joule.2020.09.002.

[57] Nathanael, A., Kannaiyan, K., ... A. K.-R. C. and 2021, undefined. Global opportunities and challenges on net-zero CO_2 emissions towards a sustainable future. pubs.rsc.org.
[58] Zhang Jiutian, Wang Zhiyong, Kang Jia-Ning, Song Xiangjing and Xu Dong. 2022. Several Key Issues for CCUS Development in China Targeting Carbon Neutrality. Springer.
[59] Environmental, M. C.-I. C. S. E. and 2021, undefined. The role of subsurface engineering in the net-zero energy transition. iopscience.iop.org doi:10.1088/1755-1315/861/7/072017.
[60] Jiutian, Z., Zhiyong, W., Jia-Ning, K., Xiangjing, S. and Dong, X. 2022. Several key issues for CCUS development in China targeting carbon neutrality. Carbon Neutrality 1.
[61] Williams, J., Jones, R., Change, M. T.-E. and C. & 2021, undefined. Observations on the Transition to a Net-zero Energy System in the United States. Elsevier.
[62] Greig, C., Journal, S. U.-T. E. and 2021, undefined. The Value of CCUS in Transitions to Net-zero Emissions. Elsevier.
[63] Compass, D. R.-S. and 2022, undefined. Progress, Opportunities and Challenges of Achieving Net-zero Emissions and 100% Renewables. Elsevier.
[64] Turner, K., Katris, A. and Race, J. 2020. The need for a Net Zero Principles Framework to support public policy at local, regional and national levels. Local Econ. 35: 627–634.
[65] Ken Oshiro, Shinichiro Fujimori, Tomoko Hasegawa, Shinichiro Asayama, Hiroto Shiraki and Kiyoshi Takahashi. 2023. Alternative, but expensive, energy transition scenario featuring carbon capture and utilization can preserve existing energy demand technologies. cell.com 6: 872–883.
[66] Ness, G., Sorbie, K. S., Al Mesmari, A. H. and Masalmeh, S. 2022. The impact of CCUS for improved oil recovery on $CaCO_3$ scaling potential of produced fluids. Society of Petroleum Engineers - SPE EuropEC - Europe Energy Conference featured at the 83rd EAGE Annual Conference and Exhibition, EURO 2022. doi:10.2118/209676-MS.
[67] Bang, G. 2021. The United States: conditions for accelerating decarbonisation in a politically divided country. Int. Environ. Agreem. 21: 43–58.
[68] Gössling, S., Tourism, D. S.-J. of S. and 2018, undefined. 2018. The Decarbonisation Impasse: Global Tourism Leaders' Views on Climate Change Mitigation. Taylor & Francis 26: 2071–2086.
[69] Hafner, M., Tagliapietra, S. and Mattei, F. 2016. The future of european gas markets: Balancing act between decarbonisation and security of supply.
[70] Governance, T. R.-E. S. and 2021, undefined. Taking the Slow Route to Decarbonisation? Developing Climate Governance for International Transport. Elsevier.
[71] Obergassel, W., Lah, O., Governance, F. R.-E. S. and 2021, undefined. Driving towards Transformation? To what Extent Does Global Climate Governance Promote Decarbonisation of Land Transport? Elsevier.
[72] Nehler, T. and Fridahl, M. 2022. Regulatory preconditions for the deployment of bioenergy with carbon capture and storage in Europe. Frontiers in Climate 4.
[73] Mandova, H., Patrizio, P., Leduc, S., ... J. K.-J. of C. and 2019, undefined. Achieving Carbon-Neutral Iron and Steelmaking in Europe Through the Deployment of Bioenergy with Carbon Capture and Storage. Elsevier.
[74] Coninck, H. De, Flach, T., Curnow, P., ... P. R.-I. J. of & 2009, undefined. The Acceptability of CO_2 Capture and Storage (CCS) in Europe: An Assessment of the Key Determining Factors: Part 1. Scientific, Technical and Economic. Elsevier.
[75] Nerlich, B. and Jaspal, R. 2013. UK media representations of carbon capture and storage. Metaphor and the Social World 3: 35–53.
[76] Rob Bellamy, Mathias Fridahl, Javier Lezaun, James Palmer, Emily Rodriguez, Adrian Lefvert et al. 2021. Incentivising Bioenergy With Carbon Capture And Storage (BECCS)

Responsibly: Comparing Stakeholder Policy Preferences in the United Kingdom and. Elsevier.

[77] Yi Huang, Qun Yi, Guo-qiang Wei, Jing-xian Kang, Wen-ying Li, Jie Feng et al. Oct 2018. CO_2 Emissions Evaluation of Biomass/Coal, with and without CO_2 Capture and Storage, in a Pulverized Fuel Combustion Power Plant in the United Kingdom. Elsevier.

[78] Karimi, F., GeoJournal, N. K. and 2017, undefined. 2017. Understanding Experts' Views and Risk Perceptions on Carbon Capture and Storage in Three European Countries. Springer 82: 185–200.

[79] Alphen, K. Van, Ruijven, J. Van, Kasa, S., Policy, M. H.-E. and 2009, undefined. The Performance of the Norwegian Carbon Dioxide, Capture and Storage Innovation System. Elsevier.

[80] Ishii, A., Change, O. L.-G. E. and 2011, undefined. Toward Policy Integration: Assessing Carbon Capture and Storage Policies in Japan and Norway. Elsevier.

[81] Alphen, K. van, Hekkert, M., Procedia, W. T.-E. and 2009, undefined. Comparing the Development and Deployment of Carbon Capture and Storage Technologies in Norway, the Netherlands, Australia, Canada and the United States–An. Elsevier.

CHAPTER 11

Carbon Capture and Storage Information Resources

*Abdulrab Abdulwahab Mahyoub Salem** and *Bhajan Lal*

11.1 Further Sources of Information

Organizations, initiatives, projects, and other predominantly online resources relating to CCS technologies.

11.1.1 International and National Organizations and Projects

International Organizations and Projects

The different international organizations for the projects include ADB, EU, CEM, MI, IPCC, GCCSRI, IPHE, CCUS, CCC, CIF, IETA, CCS, CSLF, and UNFCC where details with the website are provided in the table below.

The national organization with the projects and their organization include Kemper IGCC, TCEP, MIT, CWC, DOE, and CMI from Australia, Canada, United Kingdom, Norway, Germany, France, which are described in the table below.

Chemical Engineering Department, Center for Carbon Capture, Storage and Utilization (CCUS), Institute of Sustainable Energy, Universiti Teknologi PETRONAS, Bandar Seri Iskandar, Perak, Malaysia.
* Corresponding author: abdulrab_22011654@utp.edu.my

11.2 Resources by Technology Area

DOE -	U.S. Department of Energy (DOE) - Advanced Energy Systems Program	www.energy.gov
IEA -	IEA Clean Coal Centre	www.iea-coal.org
FutureGen	U.S. DoE FutureGen Initiative	www.fossil.energy.gov
CC Technologies -	Clean Coal Technologies, Inc	www. cleancoaltechnologiesinc.com
NETL -	National Energy Technology Laboratory (NETL) - Carbon Capture, Utilization, and Storage	www.netl.doe.gov
Coal Geology -	International Journal of Coal Geology	www.sciencedirect.com
World Coal Association	World Coal Association	www.worldcoal.org
GTC -	Gasification Technologies Council	www.gasification.org

11.2.1 Clean Coal-fired Generation

DOE	U.S. Department of Energy (DOE) - Advanced Energy Systems Program	www.energy.gov
IEA	IEA Clean Coal Centre	www.iea-coal.org
FutureGen	U.S. DoE FutureGen Initiative	www.fossil.energy.gov
CC Technologies	Clean Coal Technologies, Inc	www.cleancoaltechnologiesinc. com
NETL	National Energy Technology Laboratory (NETL) - Carbon Capture, Utilization, and Storage	www.netl.doe.gov
Coal Geology	International Journal of Coal Geology	www.sciencedirect.com
World Coal Association	World Coal Association	www.worldcoal.org
GTC	Gasification Technologies Council	www.gasification.org

11.2.2 Adsorption

IAS	International Adsorption Society	adsorption.org
PSA Plants	Pressure-swing adsorption trading Portal	www.psaplants.com
EPA	United States Environmental Protection Agency (EPA) - Adsorption	www.epa.gov
Journal of Adsorption Science and Technology	Journal of Adsorption Science and Technology	www.tandfonline.com
Adsorption.org	Knowledge sharing site for adsorption based systems and technologies	www.adsorption.org
ACS	American Chemical Society (ACS) - Division of Industrial and Engineering Chemistry	www.acs.org

11.2.3 Membranes and Molecular Sieves

NAMS	North American Membrane Society	www.nams.washington.edu
IMSTEC	International Membrane Science & Technology (IMSTEC) Conference	www.imstec.org
LANL	Los Alamos National Lab	www.lanl.gov
Membrane Separation Technology	Membrane Separation Technology Group at the University of Twente	www.utwente.nl/en/et/mst
NATCO	Cynara membrane separation systems	www.natcogroup.com
Membrane Technology Forum	Membrane Technology Forum	www.membraneforum.com
UOP	Separex membrane separation systems	www.uop.com
Membrana	Industrial membrane producer	www.membrana.com
MMS	Membrane and Molecular Separation (MMS) Research Group at MIT	web.mit.edu/mms
Praxair	Molecular sieve air separation unit	www.praxair.com

11.2.4 Cryogenic and Distillation Systems

AIChE	American Institute of Chemical Engineers (AIChE) - Distillation and Separation Division	www.aiche.org
Sulzer	Distillation column equipment	www.sulzerchemtech.com
Linde AG	Cryogenic air separation systems	www.linde.de
CSA	Cryogenic Society of America (CSA)	www.cryogenicsociety.org
Cryogenic and Distillation Research Group	Cryogenic and Distillation Research Group at the University of Florida	groberts.che.ufl.edu
Praxair	Cryogenic air separation systems	www.praxair.com
Distillation and Separation	Chemical Engineering Resources - Distillation and Separation	www.chemengonline.com

11.2.5 Mineral Carbonation

MCi	Mineral Carbonation International (MCi)	www.mci.com.au
CCT	Carbon Trap Technologies, L.P.	www.carbontrap.net
UBC	University of British Colombia, Earth, and Ocean Sciences Department	www2.ocgy.ubc.ca/research/dipple/UBC_Carbonation
Carbon8 Systems	Carbon8 Systems	carbon8.com
Frontiers in Mineral Carbonation	Frontiers in Mineral Carbonation Research	www.frontiersin.org
Carbon Mineralization Research Group	Carbon Mineralization Research Group at the University of Oslo	www.mn.uio.no
LANL	Los Alamos National Lab	www.lanl.gov
Mineral Carbonation International Research Hub	Mineral Carbonation International Research Hub	www.newcastle.edu.au

11.2.6 Geological Storage

CSIRO	Commonwealth Scientific and Industrial Research Organisation	www.csiro.au
ARC	Alberta Research Council	www.arc.ab.ca
CO2GeoNet	European Network of Excellence on Geological Storage of CO_2	www.co2geonet.com
Global CCS Institute	Global CCS Institute	www.globalccsinstitute.com
IEA	International Energy Agency (IEA) - Greenhouse Gas R&D Programme	ieaghg.org
AERI	Alberta Energy Research Institute	www.aeri.ab.ca
DOE	U.S. Department of Energy (DOE) - Carbon Storage Program	www.energy.gov/fe/
CCSA	Carbon Capture and Storage Association (CCSA)	www.ccsassociation.org
IEAGHG	IEAGHG Monitoring and Verification Network	www.ieaghg.org
MGSC	Midwest Geological Sequestration Consortium	www.sequestration.org
USGS	United States Geological Survey (USGS) - Carbon Sequestration	www.usgs.gov

11.2.7 Ocean Storage

IPCC	Intergovernmental Panel on Climate Change (IPCC) Special Report on the Ocean and Cryosphere in a Changing Climate	www.ipcc.ch/srocc
OCN	NOAA's Ocean Acidification Program	www.oceannourishment.com
Atmocean	Developing proprietary wave driven technology to enhance upwelling	www.atmocean.com
MBARI	Monterey Bay Aquarium Research Institute	www.mbari.org
PICHTR	Pacific International Center for High Technology Research, Hawaii	www.pichtr.
OCB	The Ocean Carbon and Biogeochemistry (OCB) Program	http://www.us-ocb.org/
Climos	Developing carbon offsets through ocean iron fertilization	www.climos.com
TOS	The Oceanography Society	www.tos.org

11.2.8 Terrestrial Ecosystem Storage

IPCC	Intergovernmental Panel on Climate Change (IPCC) - Land Use, Land-Use Change, and Forestry (LULUCF)	www.ipcc.ch
CASMGS	Consortium for Agricultural Soils Mitigation of Greenhouse Gases	http://agecon2.tamu.edu/people/faculty/mccarl-bruce/acs/
The Nature Conservancy	The Nature Conservancy - Forest Carbon	www.nature.org
NASA	NASA's Carbon Cycle & Ecosystems	www.carbon.nasa.gov
Woods Hole Research Center	The Woods Hole Research Center	www.woodwellclimate.org
Terrestrial Carbon Group	The Terrestrial Carbon Group	www.terrestrialcarbongroup.org
TERN	Terrestrial Ecosystem Research Network (TERN)	www.tern.org.au

11.3 CCS-related Online Journals and Newsletters

The different journals available for the information include the following

International Journal of Greenhouse Gas Control	https://www.sciencedirect.com/journal/international-journal-of-greenhouse-gas-control
Carbon Capture Journal	https://www.carboncapturejournal.com/allnews.aspx
Journal of CO_2 Utilization	https://www.sciencedirect.com/journal/journal-of-co2-utilization
The Carbon Brief Newsletter	https://www.carbonbrief.org/
IEAGHG Newsletter	https://www.ieaghg.org/newsletter
Green Car Congress	https://www.greencarcongress.com/
Biodiesel Magazine	www.biodieselmagazine.com
CCS Association Newsletter	https://www.ccsassociation.org/news-and-events/
Clean Coal Today (National Energy Technology Laboratory quarterly newsletter)	www.netl.doe.gov/technologies/coalpower/cctc/newsletter/newsletter.html
The International Journal of Greenhouse Gas Control (Elsevier)	https://www.sciencedirect.com/journal/international-journal-of-greenhouse-gas-control
NETL Carbon Sequestration Newsletter	http://listserv.netl.doe.gov/mailman/listinfo/sequestration

Index

H

I

J

K

L

M

N

O

P

R

S

T

U

W